W9-ADY-637

COLLEGE LIBRARY
SUFFOLK UNIVERSITY
BOSTON, MASS.

An Introduction to Linear Regression and Correlation

A Series of Books in Psychology

An Introduction to Linear Regression and Correlation

Allen L. Edwards
UNIVERSITY OF WASHINGTON

W. H. Freeman and Company
San Francisco

Library of Congress Cataloging in Publication Data

Edwards, Allen Louis.
An introduction to linear regression and correlation.

Includes index.
1. Regression analysis. 2. Correlation (Statistics)
I. Title.
QA278.2.E3 519.5′36 75-38811
ISBN 0-7167-0562-1
ISBN 0-7167-0561-3 pbk.

Printed in the United States of America

10 9 8 7 6 5 4 3 2 1

For Anthony and David

Contents

Preface

This book was written for students of one of the behavioral sciences, psychology, but students of the other behavioral sciences may also find it of interest. What I have tried to do in the book is to provide the student with a more detailed and systematic treatment of linear regression and correlation than that ordinarily given in either a first or a second course in applied statistics for students of psychology. Correlational techniques are of importance to the student of individual differences, and regression analysis is important to the general experimentalist. In this book I attempt to show both the similarities and the differences between these two methods of data analysis.

The book has been written at a level that can be understood by any student with a working knowledge of elementary algebra. I have *not* assumed that the reader has already been exposed to a first course in statistics. Many of the topics traditionally covered in the first course are not essential to an understanding of linear regression and correlation, and those that are essential have, I believe, been briefly but adequately covered in this book. Consequently, the book may be used as a text in either a first or a second course in statistics for psychology students. If students are exposed only to a single course in statistics, one in which a more traditional book is used as a text, then this book might be considered supplementary reading to provide a more detailed coverage of the topics of linear regression and correlation.

The book begins at a very elementary level with the equation for a straight line. There is a continuity in the development of each of the successive chapters. The second chapter treats some nonlinear functions that can be transformed into linear functions. Chapter 3 is concerned with values of a dependent variable Y that are subject to random variation. The student is shown how the method of least squares can be used to find a line of best fit, and the residual variance and the standard error of estimate as measures of the variation of the Y values about the line of best fit are introduced.

Chapter 4 deals with the correlation coefficient as a measure of the degree to which two variables are linearly related. The relationship of the correlation coefficient to the residual variance and standard error of estimate is explained. The coefficients of determination and nondetermination are discussed. Chapter 5 begins with an explanation of how any variable can be transformed into a standardized variable. The remainder of the chapter consists of a review of correlation and regression in terms of standardized variables. Various factors that may be related to the magnitude of the correlation coefficient are discussed in Chapter 6. In Chapter 7 the phi coefficient, the point biserial coefficient, and the rank order coefficient are shown to be merely special cases of the correlation coefficient.

Chapter 8 begins with a discussion of a model for a correlational problem. There is a brief discussion of tests of significance and of the four major distributions—the normal, t, F, and χ^2 distributions—used in making such tests. The treatment is nonmathematical and intuitive, and is at a level that can be understood by the beginning student. The t test of the null hypothesis that the population correlation is zero and Fisher's z_r transformation for the correlation coefficient are then discussed. The standard normal distribution test of the difference between two independent correlation coefficients is illustrated, along with the χ^2 test of the homogeneity of several independent values of the correlation coefficient. Chapter 9 is concerned with tests of significance for the special cases of the correlation coefficient; Chapter 10 deals with tests of significance for regression coefficients.

Coefficients for orthogonal polynomials are introduced in Chapter 11. Examples of correlation and regression of mean Y values with these coefficients are discussed. In Chapter 12 an example is given of the analysis of variance for an experiment involving equally spaced values of an independent variable. Tests of significance for the linear, quadratic, and other components of the treatment sum of squares are illustrated. Chapter 13 includes a description of the analysis of variance for an experiment in which the same subjects are tested with each of the equally spaced values of an independent variable.

The book concludes with a discussion of multiple regression and correlation. For simplicity, the numerical example involves only two X variables, but the basic principles are generalized to the case of more than two X variables.

At the end of each chapter I have provided a number of simple exercises designed to test the reader's understanding of the material covered in that chapter. Answers to all of the exercises that require calculations are given in the Answers to the Exercises section. Although the calculations are relatively simple, I strongly recommend that the reader buy and use a small electronic calculator to perform them. Excellent minicalculators with a memory can now be purchased for less than $30 and without a memory for less than $20. A minicalculator makes even difficult arithmetic a joy rather than a chore and, in addition, affords accuracy in calculations.

In some exercises I have asked for a proof. When the proof has already been given in the text, it is not repeated in the Answers to the Exercises. When the proof has not been given in the text, it is provided in the Answers to the Exercises.

Tables III and V in the Appendix have been reprinted from R. A. Fisher, *Statistical Methods for Research Workers* (14th ed.), Copyright 1972 by Hafner Press, by permission of the publisher. Table IV is reprinted from Enrico T. Federighi, Extended tables of the percentage points of Student's t distribution, *Journal of the American Statistical Association*, 1959, *54*, 683–688, by permission of the American Statistical Association. Table VII has been reprinted from George W. Snedecor and William G. Cochran, *Statistical Methods* (6th ed.), Copyright 1967 by Iowa State University Press, Ames, Iowa, by permission of the publisher. Table VIII has been reprinted from *Essentials of Trigonometry*, by D. E. Smith, W. D. Reeve, and E. L. Morss, Copyright 1928 by W. D. Reeve, and E. L. Morss, Copyright renewed 1956 by W. D. Reeve and E. L. Morss, published by Ginn and Company, by permission of the publishers.

For their careful reading of the manuscript, for completing the exercises at the end of the chapters, and for providing me with their reactions to and evaluations of the material contained in this book, I owe a special debt of gratitude to Clark Ashworth, Mary Cerreto, Randall M. Chestnut, Virginia deWolf, Donald Eismann, Kenneth Johnson, Lynda King, Nana Lowell, Patricia Pedigo, Gary Quarfoth, Francine Rose, Judith Siegel, Frances Thompson, Vicki Wilson, and Thomas Zieske.

Seattle, Washington Allen L. Edwards
August 1975

An Introduction to Linear Regression and Correlation

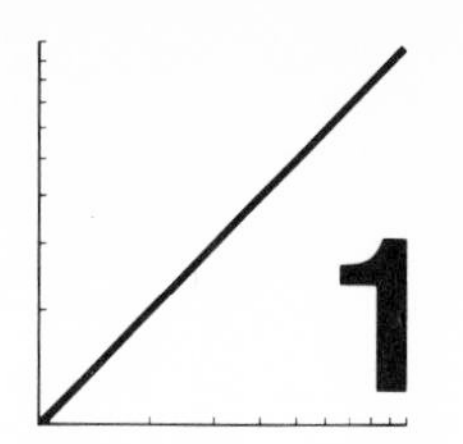

1 Linear Relationships

1.1 Introduction

Many experiments are concerned with the relationship between an independent variable X and a dependent variable Y. The values of the independent variable may represent measures of time, number of trials, varying levels of illumination, varying amounts of practice, varying dosages of a drug, different intensities of shock, different levels of reinforcement, or any other quantitative variable of experimental interest. Ordinarily, the values of the X variable *in an experiment* are selected by the experimenter and are limited in number. They are usually measured precisely and can be assumed to be without error. In general, we shall refer to the values of the X variable in an experiment as fixed in that any conclusions based on the outcome of the experiment will be limited to the particular X values investigated.

For each of the X values, one or more observations of a relevant dependent Y variable are obtained. The objective of the experiment is to determine whether the Y values (or the average Y values, if more than one observation is obtained for each value of X) are related to the X values. In this chapter we shall be concerned with the case where the Y values are linearly related to the X values. By "linearly related" we mean that if the Y values are plotted against the X values, the resulting trend of the plotted points can be represented by a straight line. If the Y values are linearly related to the X values, then we also want to determine the equation for the straight line. We may regard this equation as a rule that relates the Y values to the X values.

1.2 The Equation of a Straight Line

Consider the values of X and Y shown in Table 1.1. What is the rule that relates the values of Y to the values of X? Examination of the pairs of values will show that for each value of X, the corresponding value of Y is equal to $-.4X$. We may express this rule in the following way:

$$Y = bX \tag{1.1}$$

where $b = -.4$ is a constant that multiplies each value of X. If each value of Y in Table 1.1 were exactly equal to the corresponding value of X, then the value of b would have to be equal to 1.00. If each value of Y were numerically equal to X, but opposite in sign, then the value of b would have to be equal to -1.00.

Now examine the values of X and Y in Table 1.2. The rule or equation relating the Y values to the X values in this case has the general form

$$Y = a + bX \tag{1.2}$$

where b is again a constant that multiplies each value of X and a is a constant that is added to each of the products. For the values of X and Y given in Table 1.2, the value of b is equal to .3 and the value of a is equal to 2. Thus when $X = 10$, we have $Y = 2 + (.3)(10) = 5$. When $X = 8$, we have $Y = 2 + (.3)(8) = 4.4$.

Both (1.1) and (1.2) are equations for a straight line. For example, we could take any arbitrary constants for a and b. Then for any given set of X values we could substitute in (1.2) and obtain a set of Y values. If these values of Y are plotted against the corresponding X values, the set of plotted points will fall on a straight line.

1.3 Graph of $Y = a + bX$

Table 1.3 gives another set of X and Y values. Let us plot the Y values against the corresponding X values. The resulting graph will provide some additional insight into the nature of the constant b that multiplies each value of X as well as the nature of the constant a that is added to the

TABLE 1.1
$Y = -.4X$

X	Y
10	−4.0
9	−3.6
8	−3.2
7	−2.8
6	−2.4
5	−2.0
4	−1.6
3	−1.2
2	− .8
1	− .4

TABLE 1.2
$Y = 2 + .3X$

X	Y
10	5.0
9	4.7
8	4.4
7	4.1
6	3.8
5	3.5
4	3.2
3	2.9
2	2.6
1	2.3

TABLE 1.3
$Y = 3 + .5X$

X	Y
10	8.0
9	7.5
8	7.0
7	6.5
6	6.0
5	5.5
4	5.0
3	4.5
2	4.0
1	3.5

product. In making the graph we set up two axes at right angles to each other. It is customary to let the horizontal axis represent the independent or X variable and the vertical axis represent the dependent or Y variable. We need not begin our scale on the X and Y axes at zero. We may begin with any convenient values that permit us to plot the smallest values of X and Y. In Figure 1.1, for example, we begin the X scale with 0 and the Y scale with 2.0. Nor is it necessary that the X and Y scales be expressed in the same units, as they are in Figure 1.1.

You will recall that a pair of (X,Y) values represents the coordinates of a point. To find the point on the graph corresponding to (10, 8.0), we go out the X axis to 10 and imagine a line perpendicular to the X axis erected at this point. We now go up the Y axis to 8.0 and imagine another line perpendicular to the Y axis erected at this point. The intersection of the two perpendiculars will be the point (10, 8.0) on the graph. It is obviously not necessary to draw the perpendiculars in order to plot a set of points.

1.4 The Slope and Intercept of a Straight Line

It is clear that the points plotted in Figure 1.1 fall along a straight line. The equation of this line, as given by (1.2), is

$$Y = a + bX$$

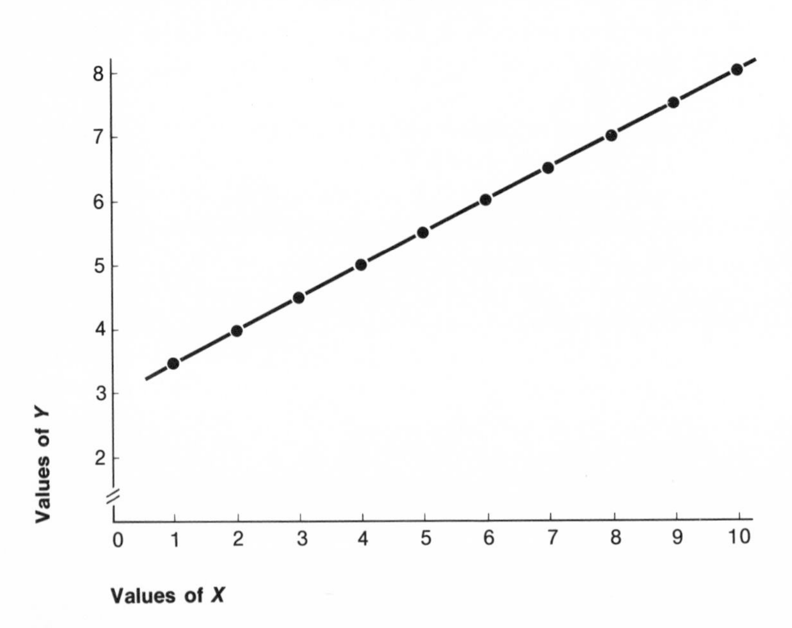

Figure 1.1 Plot of the (X,Y) values given in Table 1.3.

What is the nature of the multiplying constant b? Note, for example, that as we move from 7 to 8 on the X scale, the corresponding increase on the Y scale is from 6.5 to 7.0. An increase of one unit in X, in other words, results in .5 of a unit increase in Y. The constant b is simply the rate at which Y changes with unit change in X.

The value of b can be determined directly from Figure 1.1. For example, if we take any two points on the line with coordinates (X_1,Y_1) and (X_2,Y_2), then

$$b = \frac{Y_2 - Y_1}{X_2 - X_1} \tag{1.3}$$

Substituting in (1.3) the coordinates (2, 4.0) and (3, 4.5), we have

$$b = \frac{4.5 - 4.0}{3 - 2} = .5$$

In geometry (1.3) is known as a particular form of the equation of a straight line, and the value of b is called the slope of the straight line.

The nature of the additive constant a in (1.2) can readily be determined by setting X equal to zero. The value of a must then be the value of Y when X is equal to zero. If the straight line in Figure 1.1 were to be extended downward, we would see that the line would intersect the Y axis at the point $(0,a)$. The number a is called the Y-intercept of the line. In our example, it is easy to see that the value of a is equal to 3. If a straight line passed through the point (0,0), then a would be equal to zero, and the equation of the straight line would be $Y = bX$.

1.5 Positive and Negative Relationships

We may conclude that if the relationship between two variables is linear, then the values of a and b can be determined by plotting the values and finding the Y-intercept and the slope of the line, respectively. A single equation may then be written that will express the nature of the relationship. When the value of b is positive, the relationship is also described as positive; that is, an increase in X is accompanied by an increase in Y and a decrease in X is accompanied by a decrease in Y. When the value of b is negative, the relationship is also described as negative. A negative relationship means that an increase in X is accompanied by a decrease in Y, and a decrease in X is accompanied by an increase in Y. When two variables are positively related, the line representing the relationship will extend from the lower left of the graph to the upper right, and the slope of the line will be positive. When the relationship is negative, the line will extend from the upper left of the graph to the lower right, and the slope of the line will be negative.

Exercises

1.1. Find the values of a and b in the equation $Y = a + bX$ for the following paired (X,Y) values:

X	Y
1	2.2
2	2.6
3	3.0
4	3.4
5	3.8

1.2. Find the values of a and b in the equation $Y = a + bX$ for the following paired (X,Y) values:

X	Y
1	−5.4
2	−5.8
3	−6.2
4	−6.6
5	−7.0

1.3. Find the values of a and b in the equation $Y = a + bX$ for the following paired (X,Y) values:

X	Y
2	10.6
4	11.2
6	11.8
8	12.4
10	13.0

1.4. Find the values of a and b in the equation $Y = a + bX$ for the following paired (X,Y) values:

X	Y
1.0	4.8
1.5	4.7
2.0	4.6
2.5	4.5
3.0	4.4

1.5. Find the values of a and b in the equation $Y = a + bX$ for the following paired (X,Y) values:

X	Y
20	0
16	2
10	5
6	7
0	10

1.6. Briefly explain the meaning of each of the following terms or concepts:

dependent variable	Y-intercept
independent variable	positive relationship
slope of a straight line	negative relationship

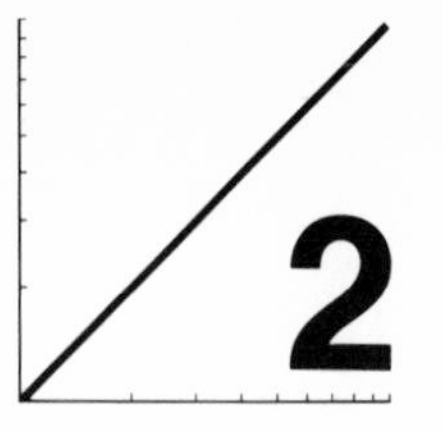

2 Some Simple Nonlinear Relationships That Can Be Transformed into Linear Relationships

2.1 The Power Curve

So far we have considered only relationships between two variables that are linear. Suppose now that when the Y values are plotted against the X values, the trend cannot be adequately represented by a straight line, that is, the relationship may be curvilinear. We would again like to find the equation of the curve representing the trend.

Sometimes a transformation of the X scale, the Y scale, or both the X and Y scales into a logarithmic scale will result in a linear relationship between the transformed values of X and Y. For example, a plot of the observed Y values against the X values may result in a curve for which the general equation is

$$Y = aX^b \tag{2.1}$$

In this instance the Y values are related to some power of the X values, and the curve is called a *power curve*. If b is negative, the curve will extend downward from the upper left to the lower right. If b is positive, the curve will extend upward from the lower left to the upper right. In the exercises at the end of this chapter, we let b take various values from -2 to 2. If you plot the values of Y obtained against the given X values in these exercises, you will gain an understanding of the form of the curve when b is integral or fractional and positive or negative.

If we take logarithms of both sides of (2.1), we obtain[1]

$$\log Y = \log a + b \log X \tag{2.2}$$

which is a linear relationship in $\log Y$ and $\log X$. This will be apparent if (2.2) is compared with (1.2), which we have already shown to be the

[1]Unless otherwise specified, all logarithms are *common logarithms* for which the base is 10. Table VIII in the Appendix is a table of common logarithms.

TABLE 2.1
$Y = aX^b$

X	Y
1.0	.20
1.5	.45
2.0	.80
2.5	1.25
3.0	1.80
3.5	2.45
4.0	3.20
4.5	4.05
5.0	5.00
5.5	6.05

equation of a straight line. Consequently, we may expect the plot of the log Y values against the log X values to be a straight line with slope equal to b and Y-intercept equal to log a.

Table 2.1 shows a set of ten paired (X,Y) values. The plot of the Y values against the X values on arithmetic or ordinary coordinate paper is shown in Figure 2.1. We would like to determine whether this curve has the general form $Y = aX^b$ and, if so, to determine the values of a and b. Figure 2.2 shows a plot of the Y values against the X values on logarithmic paper. This paper is ruled in such a way that both the X and Y axes are logarithmic scales. Plotting the original X and Y values on logarithmic paper will have the same result as if we found the logarithms of X and Y and plotted them on ordinary graph paper.

It is obvious from Figure 2.2 that a linear relationship exists between log Y and log X. To find the value of b, we take any two pairs of (X,Y) values. For example, suppose we take the paired values (4.5, 4.05) and (4.0, 3.20). Then, substituting in (2.2), we have the two equations

$$\log 4.05 = \log a + b \log 4.5$$

and

$$\log 3.20 = \log a + b \log 4.0$$

Subtracting the second equation from the first, we have

$$\log 4.05 - \log 3.20 = b(\log 4.5 - \log 4.0)$$

or

$$b = \frac{\log 4.05 - \log 3.20}{\log 4.5 - \log 4.0}$$

Substituting logarithms and solving for b, we have

$$b = \frac{.6075 - .5051}{.6532 - .6021} = \frac{.1024}{.0511} = 2$$

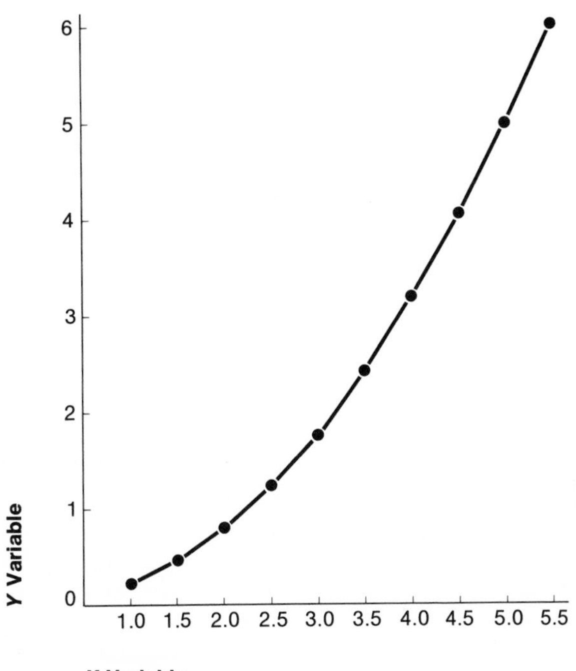

Figure 2.1 Plot of the (X,Y) values given in Table 2.1 on arithmetic coordinate paper.

Now that we know the value of b, we can easily find the value of a from any one of the paired (X,Y) values. Let us take the pair (4.0, 3.20). Then

$$\log 3.20 = \log a + 2 \log 4.0$$

or

$$\log a = \log 3.20 - 2 \log 4.0$$

Substituting logarithms in the right side of the preceding expression, we have

$$\log a = .5051 - (2)(.6021)$$

or

$$\log a = 9.3009 - 10$$

and the antilogarithm of $\log a$ is equal to .2.

We thus have the equation

$$Y = aX^b = .2X^2$$

for the data of Table 2.1.

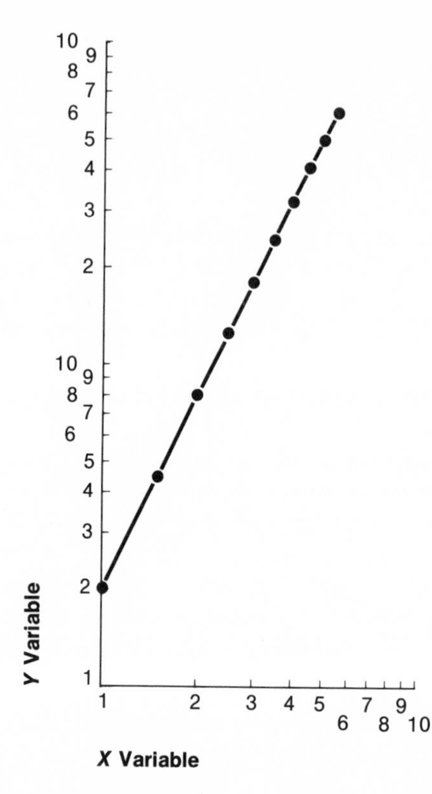

Figure 2.2 Plot of the (X,Y) values given in Table 2.1 on logarithmic paper.

2.2 The Exponential Curve

The trend of a set of plotted points may be represented by a curve for which the general equation is

$$Y = a10^{bX} \tag{2.3}$$

In this equation the independent variable appears as an exponent and the resulting curve is called an *exponential curve.*[2] If we take logarithms of both sides of (2.3), we have

---

[2]The equation may be in the form $Y = ae^{bX}$ in which e is the base of the system of *natural logarithms* and is approximately equal to 2.7183. If we take logarithms to base e of both sides of this equation, we have $\log_e Y = \log_e a + bX$, which is a linear equation in $\log_e Y$ and X. Because a table of natural logarithms has not been included in the Appendix, we may take logarithms to base 10 of both sides of the equation and obtain $\log Y = \log a + .4343bX$. This is possible because the logarithm of e to base 10 is approximately .4343. Whenever logarithms are written without a subscript, they are usually common logarithms.

$$\log Y = \log a + bX \tag{2.4}$$

which is a linear equation in log Y and the original values of X.

It is easy to determine whether the trend of a set of plotted points can be represented by a curve of the kind given by (2.3). If we plot the logarithms of Y against the values of X on ordinary arithmetic graph paper, we should obtain a straight line. It is simpler, however, to plot the original values of X and Y on semilogarithmic paper. This paper has the usual linear scale on one axis, but a logarithmic scale on the other axis. Thus, if we plot the values of X on the linear scale and the values of Y on the logarithmic scale, the plotted points should fall on a straight line. This procedure is much simpler than plotting X and log Y on ordinary coordinate paper.

Table 2.2 shows a set of ten paired (X,Y) values. The plot of the Y values against the X values on ordinary coordinate paper is shown in Figure 2.3. In Figure 2.4 the same values are plotted on semilogarithmic paper, and it is apparent that the values of Y all fall on a straight line.

To determine the value of the constant b, we take any two pairs of (X,Y) values. Suppose, for example, we take the values (2.0, 6.00) and (3.0, 3.70). Then, substituting in (2.4), we have the two equations

$$\log 6.00 = \log a + b(2.0)$$

and

$$\log 3.70 = \log a + b(3.0)$$

Subtracting the second equation from the first, we have

$$\log 6.00 - \log 3.70 = b(2.0 - 3.0)$$

Substituting logarithms and solving for b, we obtain

$$b = \frac{.7782 - .5682}{2.0 - 3.0} = -.21$$

TABLE 2.2
$Y = a10^{bX}$

X	Y
1.0	9.73
1.5	7.64
2.0	6.00
2.5	4.71
3.0	3.70
3.5	2.91
4.0	2.28
4.5	1.75
5.0	1.41
5.5	1.12

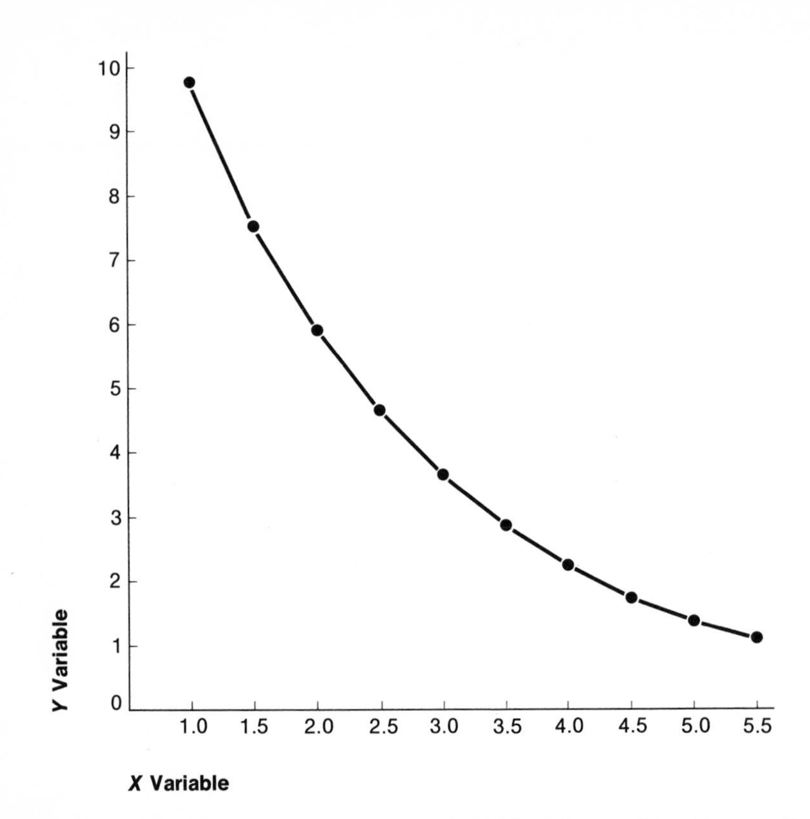

Figure 2.3 Plot of the (X,Y) values given in Table 2.2 on arithmetic coordinate paper.

Then, for the paired values (2.0, 6.00), we have

$$\log 6.00 = \log a + (-.21)(2.0)$$

or

$$\log a = .7782 + .42 = 1.1982$$

and the antilogarithm of 1.1982 is equal to 15.78.

Thus for the data of Table 2.2, we have

$$Y = a10^{bX} = (15.78)(10)^{-.21X}$$

2.3 The Logarithmic Curve

We consider one additional case, in which

$$Y = a + b \log X \tag{2.5}$$

is a linear equation in Y and $\log X$. Consequently, if (2.5) applies, and if we plot the Y values against $\log X$ on ordinary coordinate paper, the

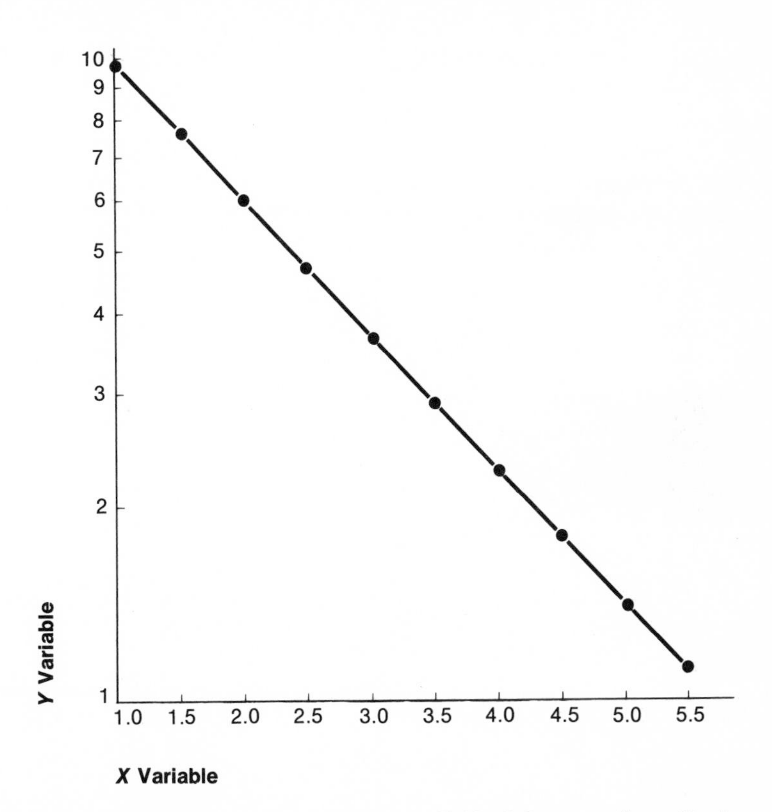

Figure 2.4 Plot of the (X,Y) values given in Table 2.2 on semilogarithmic paper.

graph should be a straight line. It is again simpler, however, to plot the original values of X and Y on semilogarithmic paper. To do so, we use the logarithmic scale for the X axis and the linear scale for the Y axis.

Table 2.3 gives a set of ten paired (X,Y) values. The plot of the Y values against the X values on ordinary coordinate paper is shown in

TABLE 2.3
$Y = a + b \log X$

X	Y
1.0	2.000
1.5	3.761
2.0	5.010
2.5	5.979
3.0	6.771
3.5	7.441
4.0	8.021
4.5	8.532
5.0	8.990
5.5	9.404

Figure 2.5. The plot of the Y values against the X values on semilogarithmic paper is shown in Figure 2.6, and it is obvious that in this graph the Y values fall along a straight line.

To find the value of b, we again take two sets of paired (X, Y) values. For example, taking (2.5, 5.979) and (2.0, 5.010) and substituting in (2.5), we have the two equations

$$5.979 = a + b \log 2.5$$

and

$$5.010 = a + b \log 2.0$$

Subtracting the second equation from the first, we obtain

$$5.979 - 5.010 = b(\log 2.5 - \log 2.0)$$

or

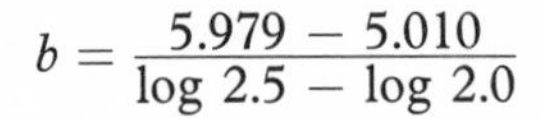

$$b = \frac{5.979 - 5.010}{\log 2.5 - \log 2.0}$$

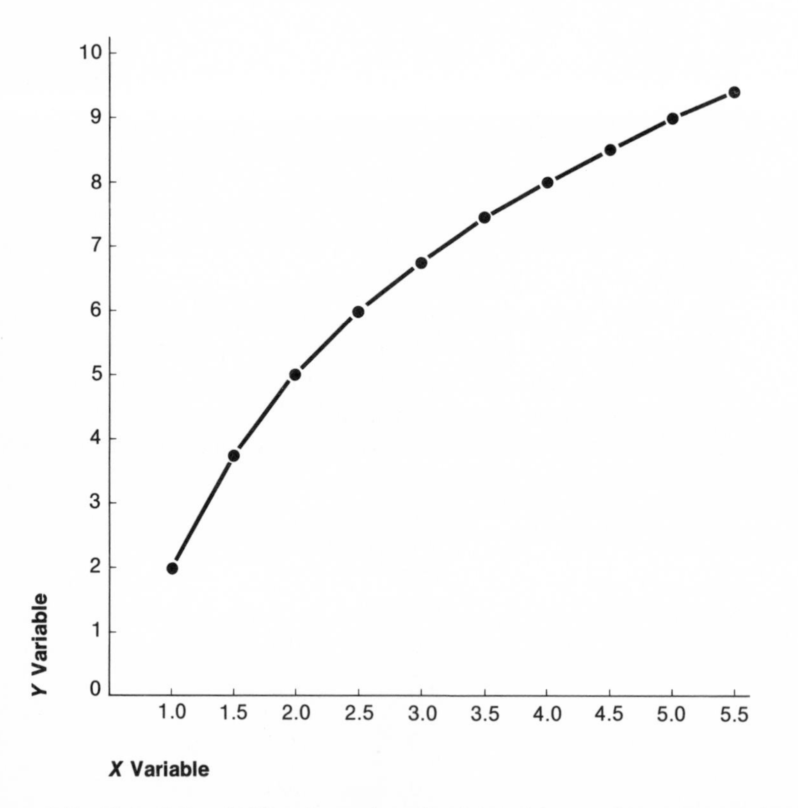

Figure 2.5 Plot of the (X, Y) values given in Table 2.3 on arithmetic coordinate paper.

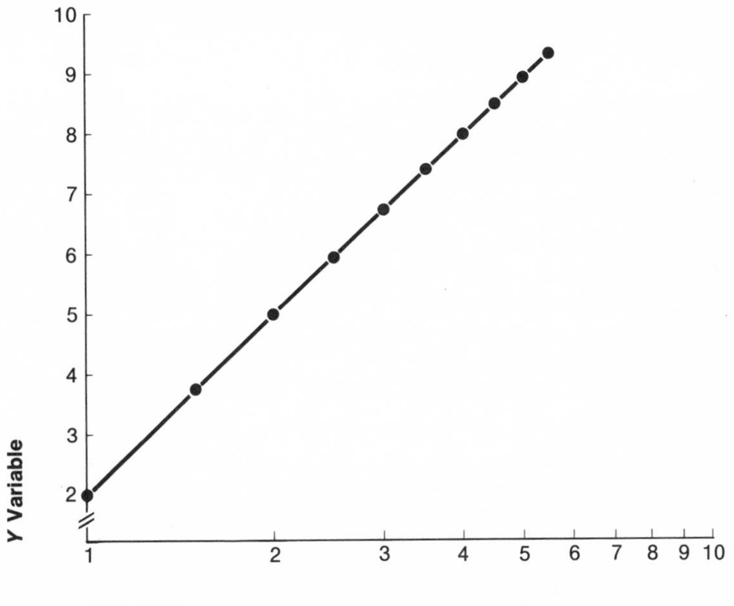

Figure 2.6 Plot of the (X, Y) values given in Table 2.3 on semilogarithmic paper.

Substituting logarithms in the preceding expression and solving for b, we obtain

$$\frac{5.979 - 5.010}{.3979 - .3010} = \frac{.9690}{.0969} = 10$$

For the value of a, we have

$$\begin{aligned} 5.010 &= a + 10 \log 2 \\ &= a + (10)(.3010) \end{aligned}$$

and

$$a = 5.010 - 3.010 = 2$$

For the data of Table 2.3, we then have

$$Y = a + b \log X = 2 + 10 \log X$$

Exercises

2.1. Assume that $Y = aX^b$, where $a = 2$ and $b = 2$. (a) Find the values of Y for the following values of X: 1.0, 1.2, 1.4, 1.6, 1.8, 2.0. (b) Plot the Y values against the X values on ordinary coordinate paper. (c) Plot the Y values against the X values on logarithmic paper.

2.2. Assume that $Y = aX^b$ where $a = 2$ and $b = -.5$. (a) Find the values of Y for the following values of X: 1, 4, 9, 16, 25, 36. (b) Plot the Y values against the X values on ordinary coordinate paper. (c) Plot the Y values against the X values on logarithmic paper.

2.3. Assume that $Y = aX^b$, where $a = 2$ and $b = .5$. (a) Find the values of Y for the following values of X: 1, 4, 9, 16, 25, 36. (b) Plot the Y values against the X values on ordinary coordinate paper. (c) Plot the Y values against the X values on logarithmic paper.

2.4. Assume that $Y = aX^b$, where $a = 2$ and $b = -2$. (a) Find the values of Y for the following values of X: 1.0, 1.2, 1.4, 1.6, 1.8, 2.0. (b) Plot the Y values against the X values on ordinary coordinate paper. (c) Plot the Y values against the X values on logarithmic paper.

2.5. Determine whether a curve of the form $Y = a10^{bX}$ will fit the following paired (X,Y) values:

X	Y
1.0	2.5
1.3	2.9
1.5	3.3
1.8	3.9
2.1	4.5
2.3	5.1
2.5	5.6
2.8	6.6
3.1	7.8
3.4	9.2

2.6. Determine whether a curve of the form $Y = aX^b$ will fit the following paired (X, Y) values:

X	Y
1.5	7.0
2.5	5.7
4.0	4.7
6.0	3.9
15.0	2.7
30.0	2.0
50.0	1.6
70.0	1.4

2.7. Determine whether a curve of the form $Y = a + b \log X$ will fit the following paired (X, Y) values:

X	Y
1.2	2.2
1.5	2.5
1.7	2.6
2.0	2.8
3.0	3.2
4.2	3.6
7.0	4.2
10.0	4.6

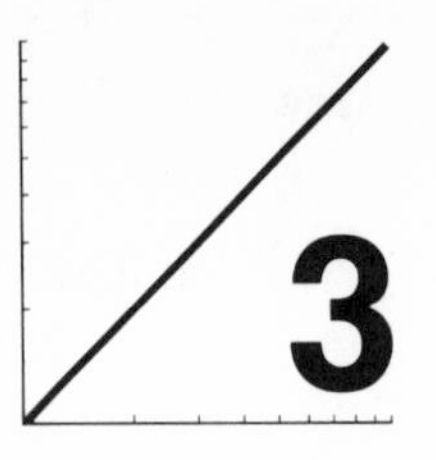

The Regression Line of *Y* on *X*

3.1 Introduction

When a set of plotted points corresponding to the values of a Y variable and an X variable all fall precisely on a straight line, with nonzero slope, so that no point deviates from the line, the relationship between the two variables is said to be *perfect*. This means that every observed value of Y will be given exactly by $Y = a + bX$. Although, as we have pointed out in the previous chapters, the values of the independent variable X selected by an experimenter may be assumed to be measured precisely and are fixed, this will ordinarily not be true of the observed values of the dependent variable Y. When the Y values obtained in an experiment are plotted against the corresponding X values, the trend of the plotted points may be linear, but the plotted points will not necessarily fall precisely on any line that we might draw to represent the trend.

When the values of X are fixed and when the values of Y are subject to random variation, the problem is to find a line of best fit that relates Y to X. This line is called the *regression line* of Y on X and the equation of the line is called a *regression equation*. Because the values of Y will not necessarily fall on the regression line, we make a slight change in notation and write the regression equation as

$$Y' = a + bX$$

The value of b in the regression equation is called a *regression coefficient*.

3.2 The Mean and Variance of a Variable

Table 3.1 gives a set of ten paired (X,Y) values and Figure 3.1 shows the plot of the Y values against the X values. It is obvious that the values of Y tend to increase as X increases, but it is also obvious that the values of

TABLE 3.1 A set of ten paired (X, Y) values and the values of $Y - \overline{Y}$, $(Y - \overline{Y})^2$, and $Y - Y'$

	(1) X	(2) Y	(3) $Y - \overline{Y}$	(4) $(Y - \overline{Y})^2$	(5) $Y - Y'$
	10	12	5.5	30.25	.01
	9	10	3.5	12.25	−.77
	8	11	4.5	20.25	1.45
	7	9	2.5	6.25	.67
	6	7	.5	.25	−.11
	5	5	−1.5	2.25	−.89
	4	4	−2.5	6.25	−.67
	3	2	−4.5	20.25	−1.45
	2	3	−3.5	12.25	.77
	1	2	−4.5	20.25	.99
Σ	55	65	0.0	130.50	0.00

Y will not fall on any straight line that we might draw to represent the upward trend of the points.

The horizontal line ($\overline{Y}$) in Figure 3.1 corresponds to the mean of the Y values. The *mean* value of a variable is defined as the sum of the observed values divided by the number of values and is ordinarily represented by a capital letter with a bar over it. For the mean of the Y values we have

$$\overline{Y} = \frac{\Sigma Y}{n} \tag{3.1}$$

where ΣY indicates that we are to sum the n values of Y. For the mean of the Y values in Table 3.1, we have

$$\overline{Y} = \frac{2 + 3 + 2 + 4 + \cdots + 12}{10} = 6.5$$

Similarly, for the mean of the X values, we have

$$\overline{X} = \frac{1 + 2 + 3 + 4 + \cdots + 10}{10} = 5.5$$

In Figure 3.1 vertical lines connect each plotted point and the mean (6.5) of the Y values. Each of these vertical lines represents a deviation of an observed value of Y from the mean of the Y values or the magnitude of $Y - \overline{Y}$. The values of these deviations are shown in column (3) of Table 3.1 and we note that the algebraic sum of the deviations is equal to zero, that is,

$$\Sigma(Y - \overline{Y}) = 0$$

This will always be true, regardless of the actual values of Y. For example, we will always have n values of $Y - \overline{Y}$ and, consequently, when we sum the n values, we obtain

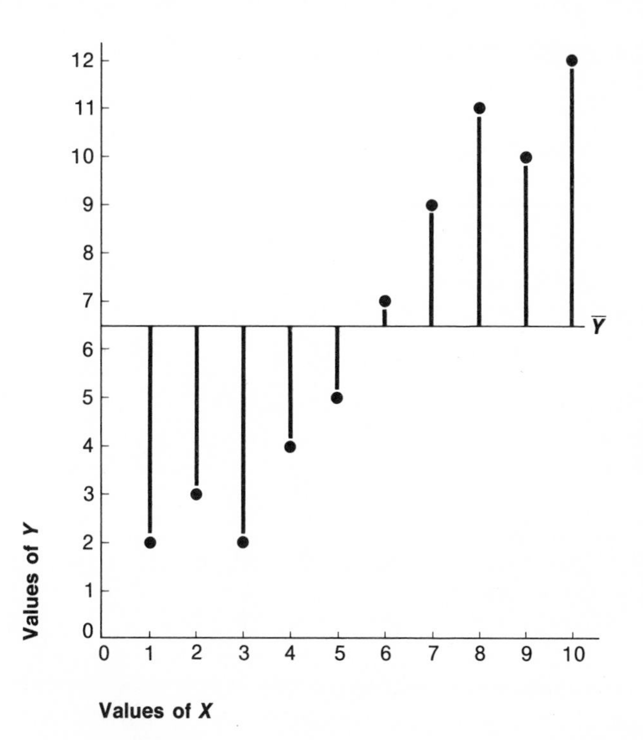

Figure 3.1 A plot of the (X, Y) values given in Table 3.1 showing the deviations of the Y values from the mean of the Y values.

$$\Sigma(Y - \overline{Y}) = \Sigma Y - n\overline{Y} = \Sigma Y - \Sigma Y = 0$$

because $\overline{Y}$ is a constant and will be subtracted n times. By definition, $\overline{Y} = \Sigma Y/n$ and, therefore, $n\overline{Y}$ will be equal to ΣY.

The squares of the deviations, $(Y - \overline{Y})^2$, are given in column (4) of Table 3.1, and for the sum of the squared deviations we have

$$\Sigma(Y - \overline{Y})^2 = 130.50$$

It can be proved,[1] for any variable Y, that $\Sigma(Y - \overline{Y})^2$ will always be smaller than $\Sigma(Y - c)^2$, where c is any constant such that c is not equal to $\overline{Y}$. For example, if we were to subtract the constant $c = 6.0$ from each

[1]A simple proof requires a knowledge of the rules of differentiation as taught in the first course of the calculus. We want to find the value of the constant c that minimizes $\Sigma(Y - c)^2$. Expanding this expression we have

$$\Sigma Y^2 - 2c\Sigma Y + nc^2$$

Differentiating the latter expression with respect to c and setting the derivative equal to zero, we obtain $c = \overline{Y}$.

value of Y in Table 3.1, then we would find that $\Sigma(Y - 6.0)^2$ is larger than $\Sigma(Y - 6.5)^2$ because $c = 6.0$ is not equal to $\overline{Y} = 6.5$.

If we divide the sum of squared deviations from the mean by $n - 1$, we obtain a measure known as the *variance*. The variance is ordinarily represented by s^2, and for the Y variable we have

$$s_Y^2 = \frac{\Sigma(Y - \overline{Y})^2}{n - 1} \tag{3.2}$$

or, for the Y values shown in Table 3.1,

$$s_Y^2 = \frac{130.50}{10 - 1} = 14.50$$

The square root of the variance is called the *standard deviation*. Thus

$$s_Y = \sqrt{\frac{\Sigma(Y - \overline{Y})^2}{n - 1}} \tag{3.3}$$

and for the Y values given in Table 3.1 we have

$$s_Y = \sqrt{\frac{130.50}{10 - 1}} = 3.81$$

The variance and standard deviation are both measures of the variability of the Y values about the mean of the Y values. When the variance and standard deviation are small, the values of Y will tend to have small deviations from the mean; when the variance and standard deviation are large, the values of Y will tend to have large deviations from the mean.

3.3 Finding the Values of *a* and *b* in the Regression Equation

Figure 3.2 is another plot of the Y values of Table 3.1 against the X values. This figure shows the horizontal line through the mean Y value (6.5) and also the vertical line through the mean X value (5.5). These two lines will obviously intersect or cross at the point with coordinates (5.5, 6.5). This point is shown in Figure 3.2 as a small open circle and is labeled B. Now suppose we rotate the horizontal line represented by $\overline{Y}$ counterclockwise about the point B until we come to the dashed line Y'. This line, as drawn in Figure 3.2, is the regression line of Y on X. The equation of the regression line, as we shall shortly see, is

$$Y' = a + bX \tag{3.4}$$

where

$$a = \overline{Y} - b\overline{X}$$

and

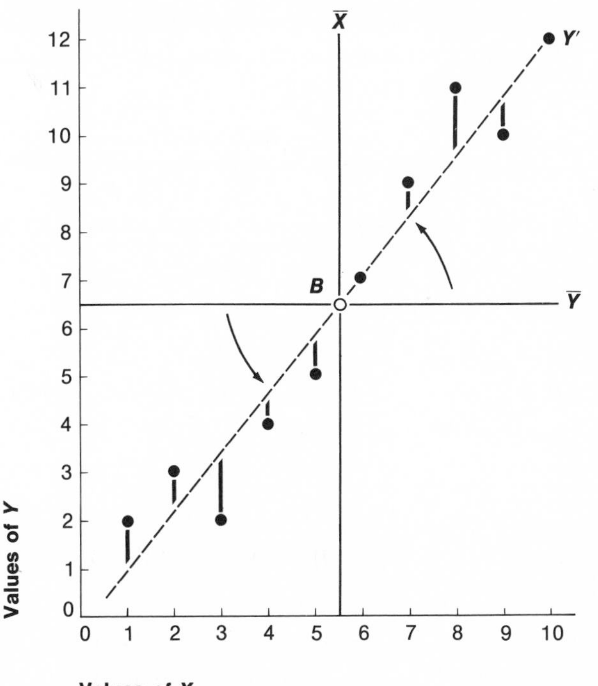

Figure 3.2 Plot of the (X,Y) values given in Table 3.1 and the line of best fit.

$$b = \frac{\Sigma(X - \overline{X})(Y - \overline{Y})}{\Sigma(X - \overline{X})^2}$$

Y', as given by (3.4), will no longer necessarily be equal to the observed value of Y corresponding to an observed value of X, if the relationship between Y and X is not perfect. We can, however, regard the value of Y' as a prediction of the observed value of Y for the corresponding value of X. Then an error of prediction will be given by

$$Y - Y' = Y - (a + bX) \tag{3.5}$$

These errors of prediction or deviations of Y from Y' are shown by the vertical lines in Figure 3.2, and the values of $Y - Y'$ are given in column (5) of Table 3.1. Note that $\Sigma(Y - Y') = 0$. We shall show why this is true later in this chapter.

The line Y' drawn in Figure 3.2 was determined by the *method of least squares*. This criterion demands that the values of a and b in equation (3.4) be determined in such a way that the *residual sum of squares*

$$\Sigma(Y - Y')^2 = \Sigma[Y - (a + bX)]^2 \tag{3.6}$$

will be a minimum. It can be shown that the values of a and b in (3.4) that will make the residual sum of squares $\Sigma(Y - Y')^2$ a minimum must satisfy the following two equations:[2]

$$\Sigma Y = na + b\Sigma X \tag{3.7}$$

$$\Sigma XY = a\Sigma X + b\Sigma X^2 \tag{3.8}$$

If we divide both sides of (3.7) by n and solve for a, we have

$$a = \overline{Y} - b\overline{X} \tag{3.9}$$

where

$$\overline{Y} = \frac{\Sigma Y}{n}$$

is the mean of the Y values and

$$\overline{X} = \frac{\Sigma X}{n}$$

is the mean of the X values. If we now substitute $\overline{Y} - b\overline{X}$ for a in (3.8), we have

$$\begin{aligned} \Sigma XY &= (\overline{Y} - b\overline{X})\Sigma X + b\Sigma X^2 \\ &= n\overline{X}\overline{Y} - bn\overline{X}^2 + b\Sigma X^2 \\ &= n\overline{X}\overline{Y} + b(\Sigma X^2 - n\overline{X}^2) \end{aligned}$$

Solving for b, we have

$$b = \frac{\Sigma XY - n\overline{X}\overline{Y}}{\Sigma X^2 - n\overline{X}^2}$$

or

$$b = \frac{\Sigma XY - (\Sigma X)(\Sigma Y)/n}{\Sigma X^2 - (\Sigma X)^2/n} \tag{3.10}$$

and b, as defined by (3.10), is called the regression coefficient.

The necessary values for calculating b are given in Table 3.2. Substituting these values in (3.10), we have

$$b = \frac{458 - (55)(65)/10}{385 - (55)^2/10} = \frac{100.5}{82.5} = 1.22$$

---

[2]Again, a simple proof requires a knowledge of the rules of differentiation. By expanding the right side of (3.6), we obtain

$$\Sigma Y^2 - 2a\Sigma Y - 2b\Sigma XY + na^2 + 2ab\Sigma X + b^2\Sigma X^2$$

Differentiating this expression with respect to a and with respect to b and setting the two derivatives equal to zero, we obtain (3.7) and (3.8).

Because $\overline{X} = 55/10 = 5.5$ and $\overline{Y} = 65/10 = 6.5$, and we have just found that $b = 1.22$, we may substitute in (3.9) and find

$$a = 6.5 - (1.22)(5.5) = -.21$$

The regression equation for the data of Table 3.2 will then be

$$Y' = -.21 + 1.22X$$

Note now that if we predict a value of Y corresponding to the mean of the X values, we obtain

$$Y' = -.21 + (1.22)(5.5) = 6.5$$

which is equal to the mean of the Y values. The regression line will, therefore, pass through the point established by the means of the X and Y values or, in other words, the point with coordinates $(\overline{X}, \overline{Y})$. This will always be true for any linear regression line fitted by the method of least squares.

The predicted value of Y when X is equal to 3 will be

$$Y' = -.21 + (1.22)(3) = 3.45$$

and when X is equal to 9, the predicted value of Y will be

$$Y' = -.21 + (1.22)(9) = 10.77$$

The regression line will therefore pass through the points with coordinates (5.5, 6.5), (3, 3.45), and (9, 10.77). If we draw a line through these points this will be the regression line of Y on X, as shown in Figure 3.2. The predicted values of Y' for each of the other X values are given in column (6) of Table 3.2; these points all fall on the regression line shown in Figure 3.2.

TABLE 3.2 Finding a line of best fit

	(1) X	(2) Y	(3) X^2	(4) Y^2	(5) XY	(6) Y'	(7) $Y - Y'$	(8) $(Y - Y')^2$
	10	12	100	144	120	11.99	.01	.0001
	9	10	81	100	90	10.77	−.77	.5929
	8	11	64	121	88	9.55	1.45	2.1025
	7	9	49	81	63	8.33	.67	.4489
	6	7	36	49	42	7.11	−.11	.0121
	5	5	25	25	25	5.89	−.89	.7921
	4	4	16	16	16	4.67	−.67	.4489
	3	2	9	4	6	3.45	−1.45	2.1025
	2	3	4	9	6	2.23	.77	.5929
	1	2	1	1	2	1.01	.99	.9801
Σ	55	65	385	553	458	65.00	0.00	8.0730

3.4 The Covariance

The numerator of (3.10) is equal to the sum of the products of the deviations of the paired X and Y values from their respective means. Thus, we have

$$\begin{aligned}\Sigma(X - \overline{X})(Y - \overline{Y}) &= \Sigma XY - \overline{Y}\Sigma X - \overline{X}\Sigma Y + n\overline{X}\overline{Y} \\ &= \Sigma XY - n\overline{X}\overline{Y} - n\overline{X}\overline{Y} + n\overline{X}\overline{Y} \\ &= \Sigma XY - n\overline{X}\overline{Y}\end{aligned}$$

or

$$\Sigma(X - \overline{X})(Y - \overline{Y}) = \Sigma XY - \frac{(\Sigma X)(\Sigma Y)}{n} \tag{3.11}$$

It will be convenient to let $x = X - \overline{X}$ and $y = Y - \overline{Y}$. Then

$$\begin{aligned}\Sigma xy &= \Sigma(X - \overline{X})(Y - \overline{Y}) \\ &= \Sigma XY - \frac{(\Sigma X)(\Sigma Y)}{n}\end{aligned}$$

The sum of the products of the deviations of the paired X and Y values from their respective means, when divided by $n - 1$, is called the *covariance*, c_{XY}. Thus

$$c_{XY} = \frac{\Sigma xy}{n - 1} \tag{3.12}$$

and if c_{XY} is equal to zero, then b will also be equal to zero.

The denominator of (3.10) is equal to the sum of the squared deviations of the X values from the mean of the X values. Thus

$$\begin{aligned}\Sigma x^2 &= \Sigma(X - \overline{X})^2 \\ &= \Sigma X^2 - 2\overline{X}\Sigma X + n\overline{X}^2 \\ &= \Sigma X^2 - \frac{(\Sigma X)^2}{n}\end{aligned} \tag{3.13}$$

The sum of squared deviations divided by $n - 1$, as we pointed out earlier, is called the variance and is represented by s^2. Therefore, the variance of the X values is

$$s_X^2 = \frac{\Sigma x^2}{n - 1} = \frac{\Sigma(X - \overline{X})^2}{n - 1}$$

We thus see that the regression coefficient, as defined by (3.10), can also be written as

$$b = \frac{c_{XY}}{s_X^2} = \frac{\Sigma xy/(n - 1)}{\Sigma x^2/(n - 1)} = \frac{\Sigma(X - \overline{X})(Y - \overline{Y})}{\Sigma(X - \overline{X})^2} \tag{3.14}$$

3.5 The Residual Sum of Squares

The residual sum of squares or the sum of the squared errors of prediction as given by (3.6) is a measure of the variation of the Y values about the regression line. Let us see if we can gain some additional insight into the nature of this sum of squares. By definition,

$$Y' = a + bX$$

Substituting an identity for a, as given by (3.9), we have

$$Y' = \overline{Y} - b\overline{X} + bX$$

Summing over the n values, we have

$$\Sigma Y' = n\overline{Y} - bn\overline{X} + b\Sigma X$$

Because $n\overline{X} = \Sigma X$, the last two terms on the right cancel, and we have

$$\Sigma Y' = \Sigma Y$$

The sum of the predicted values, $\Sigma Y'$, is thus equal to the sum of the observed values, ΣY, and the mean of the predicted values must therefore be equal to the mean of the observed values of Y. We see that this is true for the data of Table 3.2, where $\Sigma Y' = 65.0$ and $\Sigma Y = 65.0$. It also follows that the algebraic sum of the deviations of the observed values from the predicted values must be equal to zero. Thus

$$\Sigma(Y - Y') = \Sigma Y - \Sigma Y' = 0$$

because we have just shown that $\Sigma Y' = \Sigma Y$.

We have just shown that a predicted value Y' can be written $Y' = \overline{Y} - b\overline{X} + bX$. Rearranging the last two terms, we have

$$\begin{aligned} Y' &= \overline{Y} + bX - b\overline{X} \\ &= \overline{Y} + b(X - \overline{X}) \end{aligned}$$

Then an error of prediction will also be given by

$$Y - Y' = (Y - \overline{Y}) - b(X - \overline{X}) \tag{3.15}$$

The errors of prediction, $Y - Y'$, for each of the X values are given in column (7) of Table 3.2 and the squares of the errors are given in column (8). For the sum of the squared errors we have $\Sigma(Y - Y')^2 = 8.0730$.

Substituting $y = Y - \overline{Y}$ and $x = X - \overline{X}$ in (3.15), we have

$$Y - Y' = y - bx \tag{3.16}$$

Squaring and summing, we have, for the sum of the squared errors of prediction,

$$\Sigma(Y - Y')^2 = \Sigma y^2 - 2b\Sigma xy + b^2\Sigma x^2$$

But $b = \Sigma xy/\Sigma x^2$, and substituting in the preceding expression we have

$$\Sigma(Y - Y')^2 = \Sigma y^2 - 2\frac{(\Sigma xy)^2}{\Sigma x^2} + \frac{(\Sigma xy)^2}{\Sigma x^2}$$

$$= \Sigma y^2 - \frac{(\Sigma xy)^2}{\Sigma x^2} \tag{3.17}$$

Table 3.2 gives the necessary values for finding

$$\Sigma y^2 = \Sigma(Y - \overline{Y})^2 = \Sigma Y^2 - \frac{(\Sigma Y)^2}{n}$$

Thus

$$\Sigma y^2 = 553 - \frac{(65)^2}{10} = 130.5$$

Then, because we have already found that $\Sigma xy = 100.5$ and that $\Sigma x^2 = 82.5$, we find that the residual sum of squares, as given by (3.17), is

$$\Sigma(Y - Y')^2 = 130.5 - \frac{(100.5)^2}{82.5} = 8.07$$

which is equal, within rounding errors, to the sum of the squared errors of prediction given in column (8) of Table 3.2.

Now, because $\Sigma y^2 = \Sigma(Y - \overline{Y})^2$ measures the variation of the Y values about the mean of the Y values, it is obvious from (3.15) that only if the regression coefficient b is equal to zero could the residual sum of squares $\Sigma(Y - Y')^2$ be equal to $\Sigma(Y - \overline{Y})^2$. Then we would know that there is no tendency for the Y values to be linearly related to the X values.

On the other hand, if a linear relationship between Y and X does exist, regardless of whether it is positive or negative, $\Sigma(Y - Y')^2$ will be smaller than $\Sigma(Y - \overline{Y})^2$. When the relationship is negative, then $\Sigma xy = \Sigma(X - \overline{X})(Y - \overline{Y})$ will be negative and consequently the regression coefficient will also be negative. But because Σxy is squared in (3.17), the numerator of the last term will always be positive, and $\Sigma x^2 = \Sigma(X - \overline{X})^2$, being a sum of squares, is, of course, always positive. Consequently, if either a negative or a positive relationship exists between X and Y, then $\Sigma(Y - Y')^2$ will be smaller than $\Sigma(Y - \overline{Y})^2$.

When the relationship between two variables is perfect, either positive or negative, the residual sum of squares will be equal to zero and there will be no errors of prediction. If there is no linear relationship between X and Y, the residual sum of squares will be exactly equal to the sum of squared deviations of the Y values from the mean of the Y values. In this instance, the best prediction that we could make for each Y value would be the mean of the Y values, because this prediction would minimize the sum of squared errors of prediction. In other words, the sum of squared deviations from the mean, $\Sigma(Y - \overline{Y})^2$, is less than it would be from any other *single* value not equal to $\overline{Y}$.

By taking into account the relationship between Y and X, when one exists, we reduce the total variation in Y, that is, $\Sigma(Y - \overline{Y})^2$, by an amount equal to $(\Sigma xy)^2/\Sigma x^2$. The residual sum of squares measures the remaining variation in Y that cannot be accounted for by a linear relationship. Instead of measuring the variation in the Y values in terms of their deviations from the mean of the Y values, the residual sum of squares measures the variation of the Y values from the corresponding predicted values given by the regression equation.

If you will look for a moment at Figure 3.2, which shows the regression line of Y on X, you will be able to see more clearly why $\Sigma(Y - Y')^2$ will be smaller than $\Sigma(Y - \overline{Y})^2$, if X and Y are linearly related. A horizontal line has been drawn through the mean of the Y values. The vertical deviation of each plotted point from this line represents the deviation of $Y - \overline{Y}$, and the sum of these squared deviations is equal to $\Sigma(Y - \overline{Y})^2 = 130.5$. If the horizontal line through the mean of the Y values is now rotated counterclockwise about the point B, where the coordinates are $(\overline{X}, \overline{Y})$, then the sum of squared deviations from the line will become smaller and smaller until the line coincides with the regression line—line Y' in Figure 3.2. The sum of squared deviations from the regression line is $\Sigma(Y - Y')^2$, and $\Sigma(Y - Y')^2$ will be smaller than $\Sigma(Y - \overline{Y})^2$, if there is any linear relationship between X and Y. In the present example, we have $\Sigma(Y - Y')^2 = 8.07$, whereas $\Sigma(Y - \overline{Y})^2 = 130.5$. We see that $\Sigma(Y - Y')^2$ represents a considerable reduction in the sum of squared errors of prediction relative to $\Sigma(Y - \overline{Y})^2$.

It is the second variable X that makes the regression line and $\Sigma(Y - Y')^2$ meaningful. As long as the Y measures are considered alone, the best predicted value of Y for any single value of X would be the horizontal line, or the mean of the Y values. But when there is regression of Y on X, we find that different values of Y are associated with different values of X. These associated values become our predictions when we know the relationship between the two variables.

3.6 The Residual Variance and Standard Error of Estimate

If we divide the residual sum of squares by $n - 2$, we obtain a measure known as the *residual variance*. Thus

$$s_{Y.X}^2 = \frac{\Sigma(Y - Y')^2}{n - 2} = \frac{\Sigma y^2 - (\Sigma xy)^2/\Sigma x^2}{n - 2} \tag{3.18}$$

and for the data of Table 3.2, we have

$$s_{Y.X}^2 = \frac{8.07}{10 - 2} = 1.01$$

The residual variance is, as we have pointed out, a measure of the variation of the Y values about the regression line. The dot separating the

Y and X subscripts indicates that the regression line is that of Y on X; that is, that we are predicting Y values from the corresponding X values.

The square root of the residual variance is called the *standard error of estimate*. Thus

$$s_{Y.X} = \sqrt{\frac{\Sigma(Y - Y')^2}{n - 2}} \tag{3.19}$$

and for our example, we have

$$s_{Y.X} = \sqrt{\frac{8.07}{10 - 2}} = 1.00$$

Exercises

3.1. Using the method of least squares, find the values of a and b in the equation $Y' = a + bX$ for the paired (X,Y) values in the following table. Draw the regression line of Y on X.

X	Values of Y
2	3, 6
4	2, 4, 8
6	5, 7, 10
8	5, 8, 10
10	8, 12
12	5, 9, 11

3.2. (a) Find the line of best fit for the equation $Y' = a + bX$ for the following paired (X,Y) values: (6,6), (5,4), (4,5), (3,3), (2,1), (1, −1), (−1, −2), (−2, −4), (−3, −3), (−4, −5). (b) Calculate the residual variance and the standard error of estimate. (c) What are the values of $\overline{Y}$ and $\overline{X}$? (d) What are the values of s_Y^2 and s_X^2?

3.3 Prove that $\Sigma(X - \overline{X})(Y - \overline{Y}) = \Sigma XY - (\Sigma X)(\Sigma Y)/n$.

3.4. Prove that $\Sigma(Y - Y')^2 = \Sigma y^2 - (\Sigma xy)^2/\Sigma x^2$.

3.5. If b is equal to zero, will c_{XY} be equal to zero?

3.6. Briefly explain the meaning of each of the following concepts or terms:

linear relationship	regression coefficient
regression equation	covariance
residual variance	standard error of estimate
line of best fit	regression line
variance	perfect relationship
mean	standard deviation

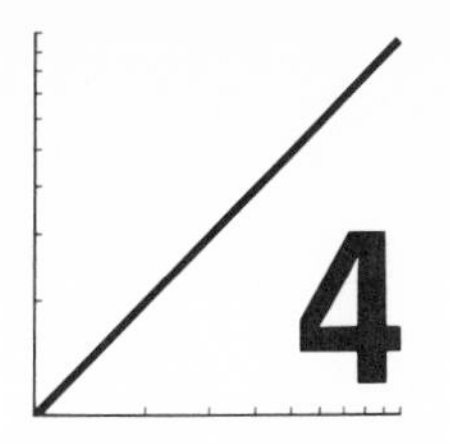

The Correlation Coefficient

4.1 Introduction

In our discussion in the first three chapters, it was assumed that we had some basis for designating one of the two variables investigated as the dependent variable Y and the other as the independent variable X. For example, if we were to measure vocabulary at selected age levels, it would seem logical to designate the vocabulary measure as the dependent variable and age as the independent variable. Vocabulary may depend on age, but it is rather difficult to imagine age as depending on vocabulary. Or suppose that one of our variables is the number of trials in a learning experiment and the other variable is the amount learned on each trial. Again it seems reasonable to regard the amount learned as depending on the number of trials rather than the number of trials as depending on the amount learned. If one of our variables is amount remembered and the other is time elapsed since learning, it seems logical to regard the amount remembered as depending on the passage of time rather than the other way around. In problems of the kind described, the experimenter ordinarily selects certain values of the independent variable X for investigation and subsequently observes the values of the dependent variable Y. He is interested in relating the values of the dependent variable Y to those of the independent variable X.

In many problems involving the relationship between two variables, however, there is no clear basis for designating one of the variables as the independent variable and the other as the dependent variable. If we have measured the heights of husbands and also the heights of their wives, which set of measures shall we designate as the dependent variable? If we have scores on a test of submissiveness and also on a test of aggressiveness, shall we consider the measure of submissiveness or the measure of aggressiveness as the dependent variable?

In problems of the kind just described, which variable we choose to designate as the dependent variable and which we choose to call the independent variable is a more or less arbitrary decision. If we arbitrarily designate one of the variables as Y and the other as X, then we may consider not only the prediction of Y values from the X values, but also the prediction of X values from the Y values. In other words, we may reverse the roles of our variables, first considering Y as a dependent variable with X as the independent variable, and then considering X as a dependent variable with Y as the independent variable.

In this chapter we shall assume there exists a population of paired (X,Y) values of the kind we have just described.[1] If we draw a sample from this population, neither the values of X nor the values of Y will be fixed. Instead, both the X values and the Y values will be subject to random variation. Under these circumstances we shall have not one but two regression lines. One will be for the regression of Y on X and the other will be for the regression of X on Y. Furthermore, we shall have two residual variances and two standard errors of estimate. There is, however, one statistic involving both variables for which we shall have but a single value. That statistic is the correlation coefficient.

4.2 The Correlation Coefficient

In discussing the correlation coefficient, we shall again restrict ourselves to the case of linear relationships. For convenience, we shall assume that one of our variables, Y, is a dependent variable and that the other variable, X, is an independent variable, so that we shall be concerned with the regression of Y on X. We will then reverse the roles of our variables and take X as the dependent variable and Y as the independent variable.

When the variables are linearly related, the *correlation coefficient* is a measure of the degree of relationship present. Consider first the existence of a perfect positive relationship between two variables, as shown in Figure 4.1(*a*). In this instance, as we shall see, the correlation coefficient is equal to 1.00. Figure 4.1(*b*) shows the plot of a set of X and Y values for which the correlation coefficient is equal to -1.00 and this is a perfect negative relationship. In Figure 4.1(*c*) the correlation between X and Y is equal to .84 and in Figure 4.1(*d*) the correlation coefficient is equal to .33.

---

[1]As an example of a population of paired (X,Y) values, consider all male students enrolled at a given university as the population of interest and let X be the height and Y be the weight of a student. For a sample of n students, we would then have n ordered pairs of (X,Y) values.

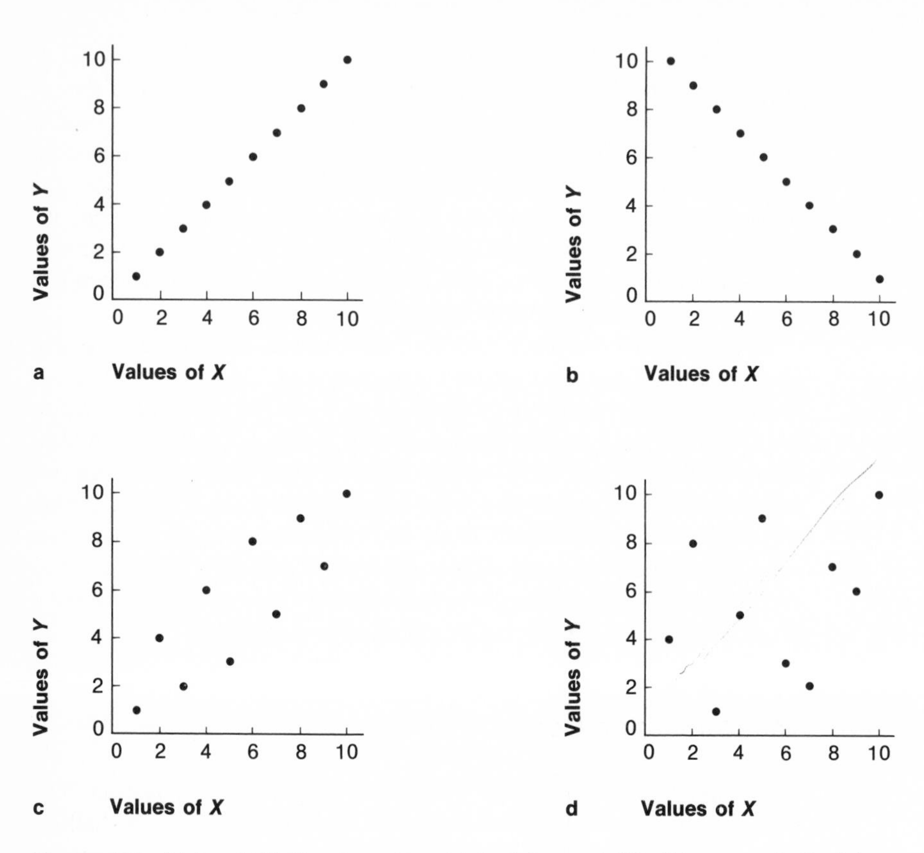

Figure 4.1 Plots of (X,Y) values for which: (*a*) $r = 1.00$, (*b*) $r = -1.00$, (*c*) $r = .84$, and (*d*) $r = .33$.

4.3 Formulas for the Correlation Coefficient

The correlation coefficient may be defined as the covariance of X and Y divided by the product of the standard deviations of the X and Y variables, or

$$r = \frac{c_{XY}}{s_X s_Y} \tag{4.1}$$

We note also that

$$c_{XY} = r s_X s_Y \tag{4.2}$$

Substituting some identities in (4.1), we also have the following equations for the correlation coefficient:

$$r = \frac{\Sigma(X - \overline{X})(Y - \overline{Y})/(n-1)}{\sqrt{\frac{\Sigma(X - \overline{X})^2}{n-1}}\sqrt{\frac{\Sigma(Y - \overline{Y})^2}{n-1}}} \tag{4.3}$$

$$r = \frac{\Sigma XY - (\Sigma X)(\Sigma Y)/n}{\sqrt{\Sigma X^2 - (\Sigma X)^2/n}\ \sqrt{\Sigma Y^2 - (\Sigma Y)^2/n}} \tag{4.4}$$

$$r = \frac{\Sigma xy}{\sqrt{\Sigma x^2}\ \sqrt{\Sigma y^2}} \tag{4.5}$$

Table 4.1 shows a set of paired (X,Y) values. Substituting the appropriate values from Table 4.1 in (4.4), we have

$$\begin{aligned} r &= \frac{274 - (32)(56)/8}{\sqrt{156 - (32)^2/8}\ \sqrt{504 - (56)^2/8}} \\ &= \frac{50}{\sqrt{28}\ \sqrt{112}} \\ &= .89 \end{aligned}$$

4.4 The Regression of Y on X

We have previously shown that the regression coefficient is equal to the ratio of the covariance of the two variables to the variance of the independent variable. Thus when Y was considered to be the dependent variable and X the independent variable, we had

$$b_Y = \frac{c_{XY}}{s_X^2} = \frac{\Sigma xy}{\Sigma x^2}$$

TABLE 4.1 A sample of $n = 8$ values of (X,Y)

	(1) X	(2) Y	(3) X^2	(4) Y^2	(5) XY	(6) Y'	(7) $Y - Y'$	(8) $(Y - Y')^2$
	7	13	49	169	91	12.37	.63	.3969
	6	9	36	81	54	10.58	−1.58	2.4964
	5	11	25	121	55	8.79	2.21	4.8841
	4	7	16	49	28	7.00	.00	.0000
	4	5	16	25	20	7.00	−2.00	4.0000
	3	7	9	49	21	5.21	1.79	3.2041
	2	1	4	1	2	3.42	−2.42	5.8564
	1	3	1	9	3	1.63	1.37	1.8769
Σ	32	56	156	504	274	56.00	0.00	22.7148

We now use the subscript Y to indicate that the regression coefficient is for Y on X. In our previous discussion of regression we were concerned only with the regression of Y on X and the subscript was not necessary.

We have already found that $\Sigma xy = 50$ and that $\Sigma x^2 = 28$ for the data of Table 4.1. The value of the regression coefficient b_Y is therefore

$$b_Y = \frac{50}{28} = 1.79$$

For the same data we have

$$\overline{Y} = \frac{56}{8} = 7$$

and

$$\overline{X} = \frac{32}{8} = 4$$

Then the regression equation for predicting Y values from the X values will be

$$Y' = a + b_Y X$$

where $a = \overline{Y} - b_Y\overline{X} = 7 - (1.79)(4) = -.16$. Substituting in the regression equation the values $a = -.16$ and $b_Y = 1.79$, we have, for the data of Table 4.1,

$$Y' = -.16 + 1.79X$$

4.5 The Residual Variance and Standard Error of Estimate

The residual sum of squares will be given by

$$\begin{aligned} \Sigma(Y - Y')^2 &= \Sigma y^2 - \frac{(\Sigma xy)^2}{\Sigma x^2} \\ &= 112 - \frac{(50)^2}{28} \\ &= 22.71 \end{aligned}$$

which is equal, within rounding errors, to the value shown in Table 4.1. Dividing the residual sum of squares by $n - 2$, we obtain the residual variance

$$s^2_{Y.X} = \frac{\Sigma(Y - Y')^2}{n - 2} = \frac{22.71}{6} = 3.7850$$

If we now take the square root of the residual variance, we obtain the standard error of estimate, or

$$s_{Y.X} = \sqrt{3.7850} = 1.9455$$

The standard error of estimate, $s_{Y.X}$, as we have pointed out previously, is a measure of the variability of the Y values about the regression line of Y on X. The standard deviation of the Y values, on the other hand, is a measure of the variation of the Y values about the mean of the Y values. In the present problem, the standard deviation of the Y values is

$$s_Y = \sqrt{\frac{112}{8 - 1}} = 4.0$$

4.6 The Regression of X on Y

If we now consider X as the dependent variable and Y as the independent variable, then the regression equation for predicting X values from Y values will be

$$X' = a + b_X Y \tag{4.6}$$

where

$$a = \overline{X} - b_X \overline{Y} \tag{4.7}$$

and

$$b_X = \frac{\Sigma xy}{\Sigma y^2} \tag{4.8}$$

For the data of Table 4.1 we have

$$b_X = \frac{50}{112} = .45$$

We also have $\overline{X} = 4$ and $\overline{Y} = 7$. Then

$$a = 4 - (.45)(7) = .85$$

and the regression equation for predicting X values from Y values will be

$$X' = .85 + .45Y$$

The residual sum of squares or sum of squared errors of prediction will be given by

$$\Sigma(X - X')^2 = \Sigma x^2 - \frac{(\Sigma xy)^2}{\Sigma y^2} \tag{4.9}$$

For the data of Table 4.1 we have

$$\Sigma(X - X')^2 = 28 - \frac{(50)^2}{112}$$

$$= 5.68$$

Then the residual variance will be given by

$$s^2_{X.Y} = \frac{\Sigma(X - X')^2}{n - 2} \tag{4.10}$$

and, in our example, we have

$$s^2_{X.Y} = \frac{5.68}{8 - 2} = .9467$$

and the standard error of estimate is thus equal to

$$s_{X.Y} = \sqrt{.9467} = .973$$

4.7 The Two Regression Lines

Figure 4.2, which is based on the data of Table 4.1, shows both the regression line of Y on X and the regression line of X on Y. The regression line of Y on X indicates the predicted change in Y as X varies. A change

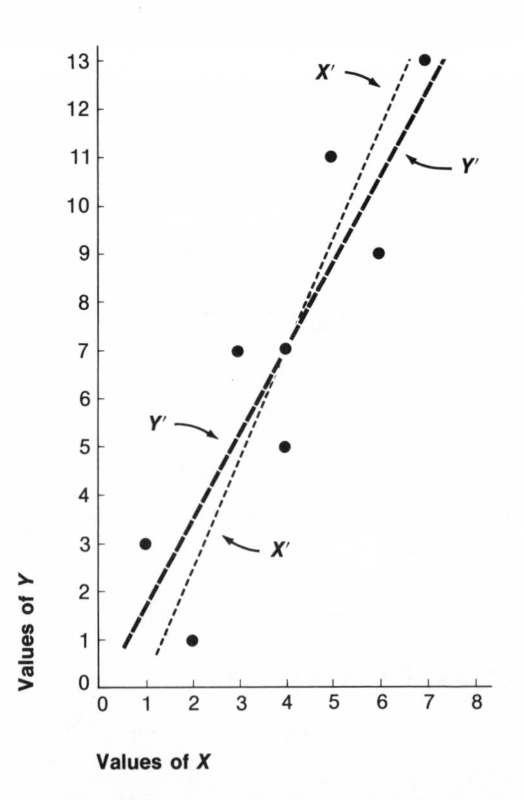

Figure 4.2 Plot of the (X, Y) values given in Table 4.1 showing the regression line (Y') of Y on X and the regression line (X') of X on Y.

of one unit in X results in a predicted change of 1.79 units in Y, and this is the value of the regression coefficient b_Y. Similarly, the regression line of X on Y indicates that with a one-unit change in the value of Y, there is a corresponding predicted change of .45 units in X, and this is the value of the regression coefficient b_X.

As the correlation coefficient approaches either of its two limiting values, -1.00 or 1.00, the two regression lines will move closer together. When r is equal to either -1.00 or 1.00, the two regression lines will coincide.

4.8 Correlation and Regression Coefficients

We see that if we consider the regression of X on Y instead of the regression of Y on X, we shall have corresponding formulas for the regression equation, regression coefficient, residual variance, and standard error of estimate. Although these formulas correspond in appearance, we should not expect them to result in identical numerical values. The only way in which these formulas could result in identical values would be if the means and the standard deviations of both the X and Y variables were identical. Let us see why this is so.

We have defined the regression coefficient of Y on X as

$$b_Y = \frac{\Sigma xy}{\Sigma x^2}$$

and we have also shown that

$$b_Y = \frac{c_{XY}}{s_X^2}$$

Recall also that

$$c_{XY} = rs_X s_Y$$

We thus have another commonly used expression for the regression coefficient of Y on X, or

$$b_Y = \frac{rs_X s_Y}{s_X^2} = r\frac{s_Y}{s_X} \tag{4.11}$$

The corresponding expression for the regression coefficient of X on Y can be obtained in the same manner and is

$$b_X = r\frac{s_X}{s_Y} \tag{4.12}$$

It is now readily apparent that only if the standard deviations of the X and Y variables are identical, will we also have $b_Y = b_X$ and, when this is true, both b_Y and b_X will be equal to r.

If we multiply the two regression coefficients, we obtain

$$b_X b_Y = r\frac{s_X}{s_Y} r\frac{s_Y}{s_X} = r^2$$

and

$$r = \pm\sqrt{r^2} = \pm\sqrt{b_X b_Y}$$

We note that both the regression coefficients must have the same sign and that the sign is determined by r. Consequently, if both b_Y and b_X are negative, then $\sqrt{r^2}$ is also negative. If both b_Y and b_X are positive, then $\sqrt{r^2}$ is positive.

4.9 Correlation and the Residual Sum of Squares

In the preceding chapter, we showed that $\Sigma(Y - Y')^2 = \Sigma y^2 - (\Sigma xy)^2/\Sigma x^2$ where $Y - Y'$ is an error of prediction resulting from the discrepancy between Y and Y' as predicted from the regression equation. If we multiply both the numerator and denominator of the last term on the right by Σy^2, we obtain

$$\Sigma(Y - Y')^2 = \Sigma y^2 - \frac{(\Sigma xy)^2(\Sigma y^2)}{(\Sigma x^2)(\Sigma y^2)}$$

But

$$r^2 = \frac{(\Sigma xy)^2}{(\Sigma x^2)(\Sigma y^2)}$$

and substituting this identity in the preceding expression, we have

$$\Sigma(Y - Y')^2 = \Sigma y^2 - r^2\Sigma y^2 \tag{4.13}$$

or

$$\Sigma(Y - Y')^2 = \Sigma y^2(1 - r^2) \tag{4.14}$$

The residual variance will then also be given by

$$s^2_{Y.X} = \frac{\Sigma y^2(1 - r^2)}{n - 2} \tag{4.15}$$

and the standard error of estimate will be given by

$$s_{Y.X} = \sqrt{\frac{\Sigma y^2(1 - r^2)}{n - 2}}$$

A similar expression can be obtained for $\Sigma(X - X')^2$ and $s^2_{X.Y}$ by substituting Σx^2 for Σy^2 in (4.14). Thus

$$\Sigma(X - X')^2 = \Sigma x^2(1 - r^2)$$

$$s_{X.Y}^2 = \frac{\Sigma x^2(1 - r^2)}{n - 2}$$

and

$$s_{X.Y} = \sqrt{\frac{\Sigma x^2(1 - r^2)}{n - 2}}$$

4.10 A Variance Interpretation of r^2 and $1 - r^2$

Suppose we regard each value of Y as being composed of two components, Y', the linearly predicted component, and $Y - Y'$, the error component. These two components will be uncorrelated with each other because the numerator of the correlation coefficient between Y' and $Y - Y'$ is equal to zero. For example, we can write the numerator of the correlation coefficient as

$$\Sigma Y'(Y - Y') - \frac{(\Sigma Y')[\Sigma(Y - Y')]}{n}$$

We know that $\Sigma(Y - Y') = 0$ and, consequently, the last term in this expression is equal to zero. We have

$$Y' = \overline{Y} + b_Y(X - \overline{X})$$

Then

$$\begin{aligned}\Sigma Y'(Y - Y') &= \Sigma[\overline{Y} + b_Y(X - \overline{X})][(Y - \overline{Y}) - b_Y(X - \overline{X})] \\ &= \Sigma(\overline{Y} + b_Y x)(y - b_Y x) \\ &= \overline{Y}\Sigma y - b_Y \overline{Y}\Sigma x + b_Y \Sigma xy - b_Y^2 \Sigma x^2\end{aligned}$$

and we know that $\Sigma y = \Sigma x = 0$. Therefore,

$$\begin{aligned}\Sigma Y'(Y - Y') &= b_Y \Sigma xy - b_Y^2 \Sigma x^2 \\ &= \frac{(\Sigma xy)^2}{\Sigma x^2} - \frac{(\Sigma xy)^2}{\Sigma x^2}\end{aligned}$$

and, consequently, the correlation coefficient between Y' and $Y - Y'$ will be equal to zero.

We also know that the variance of the error component, $Y - Y'$, is

$$s_{Y.X}^2 = \frac{\Sigma(Y - Y')^2}{n - 2} = \frac{\Sigma y^2(1 - r^2)}{n - 2}$$

For the predicted component

$$Y' = \overline{Y} + b_Y(X - \overline{X})$$

we have shown that $\Sigma Y' = \Sigma Y$, so that $\overline{Y}' = \overline{Y}$. Then, for the variance of the predicted component, we have

$$\frac{\Sigma(Y' - \overline{Y})^2}{n - 1} = \frac{b_Y^2\Sigma(X - \overline{X})^2}{n - 1}$$

$$= b_Y^2 s_X^2$$

or

$$s_{Y'}^2 = r^2 \frac{s_Y^2}{s_X^2} s_X^2 = r^2 s_Y^2 \tag{4.16}$$

Using (4.13) and (4.14), we see that

$$\Sigma y^2 = r^2 \Sigma y^2 + \Sigma y^2(1 - r^2)$$

Dividing both sides of this expression by $n - 1$, we also have

$$s_Y^2 = r^2 s_Y^2 + s_Y^2(1 - r^2)$$

But we have just shown that $s_{Y'}^2 = r^2 s_Y^2$, and we also have

$$s_{Y.X}^2 = \frac{n - 1}{n - 2} s_Y^2(1 - r^2)$$

As n increases, the fraction $(n - 1)/(n - 2)$ will approach 1 as a limit, and thus we can for all practical purposes assume that

$$s_{Y.X}^2 = s_Y^2(1 - r^2)$$

Then, if n is large, the variance of the observed Y values can be assumed to consist of the sum of the variances of the two components Y' and $Y - Y'$, or

$$s_Y^2 = s_{Y'}^2 + s_{Y.X}^2 \tag{4.17}$$

The sum of the two ratios, $s_{Y'}^2/s_Y^2$ and $s_{Y.X}^2/s_Y^2$, when n is large will be approximately equal to unity and, consequently, the two ratios can be regarded as proportions. Then the ratio $s_{Y'}^2/s_Y^2 = r^2$ is the proportion of the variance of Y that can be predicted from X, and the ratio $s_{Y.X}^2/s_Y^2 = 1 - r^2$ is the proportion of the variance of Y that is independent of X.

Under the same assumptions we made in deriving (4.17), we could show that

$$s_X^2 = s_{X'}^2 + s_{X.Y}^2$$

Then, we would also have $s_{X'}^2/s_X^2 = r^2$ as the proportion of the variance of X that can be predicted from Y and $s_{X.Y}^2/s_X^2 = 1 - r^2$ as the proportion of the variance of X that is independent of Y. Because $r_{XY}^2 = r_{YX}^2 = r^2$, the value of r^2 is often interpreted as the proportion of the variance that X and Y have in common and $1 - r^2$ is interpreted as the proportion of the variance that is not shared by X and Y. The values of r^2 and $1 - r^2$ are also sometimes referred to as the *coefficient of determination* and the *coefficient of nondetermination*, respectively.

It is obvious that the variance interpretation of r^2 and $1 - r^2$ is based on the assumption that n is large and that the relationship between X and Y is linear. For example, Y' is the linearly predicted component of Y and $Y - Y'$ represents the deviation of Y from a straight line.

Exercises

4.1. We have the following paired (X,Y) values:

X	Y
7	3
13	6
2	2
4	5
15	14
10	10
19	8
28	19
26	15
22	17

(a) Find the values of $\overline{X}$ and $\overline{Y}$. (b) Find the values of s_X^2 and s_Y^2. (c) Find the value of r. (d) Find the values of a and b in the regression equation $Y' = a + b_Y X$. (e) Find the values of a and b in the regression equation $X' = a + b_X Y$. (f) Find the values of $s_{Y.X}^2$ and $s_{X.Y}^2$.

4.2. (a) Find the correlation coefficient for the following paired (X,Y) values: (12,12), (10,13), (9,9), (8,8), (7,5), (6,6), (4,0), (2,2), (1,1), (0,3). (b) What are the values of b_Y and b_X?

4.3. Prove that $r^2 = b_X b_Y$.

4.4. Prove that the errors of prediction $(Y - Y')$ are uncorrelated with the values of $(X - \overline{X})$. *Hint*: It is sufficient to show that the numerator of the correlation coefficient is equal to zero.

4.5. Under what conditions will $r = b_Y$?

4.6. If $r = .60$, $s_Y = 8.0$, and $s_X = 10.0$, then what are the values of b_X and b_Y?

4.7. If $r = -1.00$, then what is the value of $\Sigma(Y - Y')^2$?

4.8. Is it possible for b_Y to be equal to 4.0 and for b_X to be equal to 2.0 for a given set of paired (X,Y) values? Explain why or why not.

4.9. If $b_Y = 2.0$, what is the maximum possible value for b_X?

4.10. If b_Y is equal to $-.60$, can b_X be equal to .60? Explain why or why not.

4.11. Under what conditions can b_X be equal to b_Y?

4.12. Briefly explain the meaning of each of the following terms or concepts:

correlation coefficient	coefficient of determination
residual variance	coefficient of nondetermination

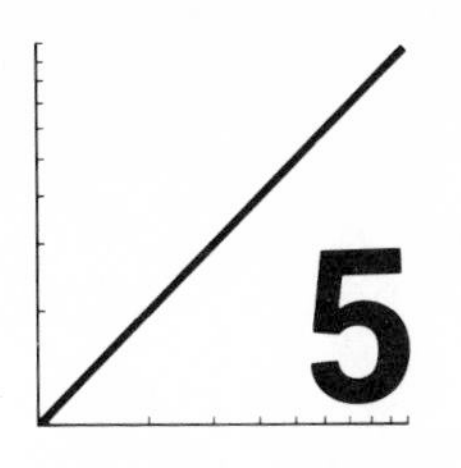

5 Correlation and Regression with Standardized Variables

5.1 Introduction

Any variable that has a mean equal to zero and a variance and standard deviation equal to one is said to be in *standardized* form. Values of such variables are often referred to as *standard scores.* Standardized variables play an important role in both psychometrics and statistics.

Any variable, regardless of its mean and standard deviation, can, by means of a linear equation, be transformed into a variable with mean equal to zero and standard deviation equal to one.

5.2 Transforming a Variable into a Standardized Variable

We have defined the mean of a variable X as

$$\overline{X} = \frac{\Sigma X}{n}$$

and the standard deviation as

$$s = \sqrt{\frac{\Sigma(X - \overline{X})^2}{n - 1}}$$

Then, for any value X_i, we define

$$z_i = \frac{X_i - \overline{X}}{s} \tag{5.1}$$

Summing each side of (5.1) over the n values, we have

$$\Sigma z = \frac{1}{s}\Sigma(X - \overline{X}) = 0$$

because the sum of the deviations from the mean is equal to zero. Thus $\bar{z}$ will be equal to zero. Then, for the variance of z, we have

$$s_z^2 = \frac{\Sigma z^2}{n-1} = \frac{\Sigma(X - \bar{X})^2}{(n-1)s^2}$$

but $\Sigma(X - \bar{X})^2 = (n-1)s^2$, and consequently $s_z^2 = 1$ and $s_z = 1$. The transformed variable z is therefore a standardized variable.

5.3 Correlation of Standardized Variables

If we have paired (X_1,X_2) values of two variables, then both variables can be put in standardized form by means of (5.1). The sum of the products of the paired (z_1,z_2) values, divided by $n - 1$, will be

$$\frac{\Sigma z_1 z_2}{n-1} = \frac{\Sigma(X_1 - \bar{X}_1)(X_2 - \bar{X}_2)}{(n-1)s_1 s_2} = \frac{c_{X_1X_2}}{s_1 s_2} = r \qquad (5.2)$$

We see that if both variables are in standardized form, the correlation coefficient is simply the sum of the products of the paired (z_1,z_2) values divided by $n - 1$.

When variables are in standardized form, the correlation coefficient is also equal to the covariance, because

$$rs_1s_2 = c_{X_1X_2}$$

and for standardized variables,

$$rs_{z_1}s_{z_2} = r$$

because both s_{z_1} and s_{z_2} are equal to one.

5.4 Regression Coefficients and Equations with Standardized Variables

We have previously shown that the regression coefficient for predicting values of X_1 from X_2 can be written as

$$b_1 = r\frac{s_1}{s_2}$$

But if both variables are in standardized form, then $s_{z_1} = s_{z_2} = 1$ and $b_1 = r$. Similarly, the regression coefficient, b_2, for predicting values of X_2 from X_1 will be equal to r. Then for the two regression equations, we have

$$z_1' = rz_2 \qquad \text{and} \qquad z_2' = rz_1$$

5.5 Residual Variance with Standardized Variables

The errors involved in predicting z_1 values from z_2 will be

$$z_1 - z_1' = z_1 - rz_2$$

Squaring and summing the errors of prediction, we have

$$\begin{aligned}\Sigma(z_1 - z_1')^2 &= \Sigma(z_1 - rz_2)^2 \\ &= \Sigma z_1^2 - 2r\Sigma z_1 z_2 + r^2 \Sigma z_2^2\end{aligned}$$

Dividing the preceding expression by $n - 1$, we obtain

$$\frac{\Sigma(z_1 - z_1')^2}{n - 1} = 1 - r^2$$

The residual variance is equal to $\Sigma(z_1 - z_1')^2$ divided by $n - 2$ and, consequently,

$$s^2_{z_1 . z_2} = \frac{n - 1}{n - 2}(1 - r^2) \tag{5.3}$$

If n is reasonably large, then $(n - 1)/(n - 2)$ will be approximately equal to one and the residual variance will be approximately equal to

$$s^2_{z_1 . z_2} = 1 - r^2 \tag{5.4}$$

which is, of course, the coefficient of nondetermination. In this case, the variance of the errors of prediction, the residual variance, is that part of the variance that X_1 and X_2 do not have in common.

Note that

$$\Sigma z_1' = r\Sigma z_2 = 0$$

and thus the mean of the predicted values will be equal to zero. Then the variance of the predicted values will be

$$\begin{aligned}s^2_{z_1'} &= \frac{\Sigma z_1'^2}{n - 1} \\ &= \frac{r^2 \Sigma z_2^2}{n - 1} \\ &= r^2 \end{aligned} \tag{5.5}$$

which is the coefficient of determination or that part of the variance that X_1 and X_2 have in common. If the variance of the predicted values, $s^2_{z_1'}$, is equal to one, then r must also be equal to one, and the z_1 values will be perfectly correlated with the z_1' values. If r is equal to zero, then all predicted values of z_1' will be equal to zero, and consequently the variance of the predicted values will also be equal to zero.

It is of some importance to observe that the errors $(z_1 - z_1')$ in predicting z_1 from z_2 are uncorrelated with the values of z_2. For example, the covariance of z_2 and $(z_1 - z_1')$ will be

$$\frac{\Sigma z_2(z_1 - z_1')}{n - 1} = \frac{\Sigma z_2(z_1 - rz_2)}{n - 1}$$

$$= \frac{\Sigma z_1 z_2 - r\Sigma z_2^2}{n - 1}$$

$$= r - r$$

$$= 0$$

and therefore the correlation between z_2 and $(z_1 - z_1')$ will also be equal to zero.

5.6 Limiting Values of the Correlation Coefficient

It can be shown that the maximum positive value of r is 1.00 and the maximum negative value is -1.00. The correlation coefficient will be equal to 1.00 only if for each and every pair of values (z_1, z_2) we have $z_1 = z_2$. Then, for each pair of values, we would have $z_1 z_2 = z^2$. Consequently,

$$r = \frac{\Sigma z_1 z_2}{n - 1} = \frac{\Sigma z^2}{n - 1} = s_z^2$$

and we know that $s_z^2 = 1.00$. If we exclude those pairs of values for which we have (0, 0), and if for each and every other pair of values the product $z_1 z_2$ is negative, and if the absolute values of z_1 and z_2 are equal, then r will be equal to -1.00.

Exercises

5.1. If $\overline{X} = 33$ and $s_X = 5.0$, and if X is transformed into a standardized variable, what will be the z value if $X = 43$?

5.2. If $c_{X_1X_2} = r$, then what do we know about s_{X_1} and s_{X_2}?

5.3. If $z = (X - \overline{X})/s_X$, prove that the mean of the z values is equal to zero and that the variance of the z values is equal to one.

5.4. Prove that if $z_1 = (X_1 - \overline{X}_1)/s_{X_1}$ and $z_2 = (X_2 - \overline{X}_2)/s_{X_2}$, then r will be equal to $\Sigma z_1 z_2/(n - 1)$.

5.5. Can the variance of the predicted values $z_1' = rz_2$ ever be greater than 1.00? Explain why or why not.

5.6. Explain why r can never be greater than 1.00 or less than -1.00.

5.7. For a sample of $n = 200$, the mean IQ is equal to 100 and the standard deviation is equal to 15.0. The IQ's are transformed into a standardized variable. Give the IQ corresponding to each of the following values of z: (a) $z = 0$, (b) $z = 1.0$, (c) $z = 2.0$, (d) $z = -1.0$, (e) $z = -2.0$.

5.8. Two variables, X_1 and X_2, are both transformed into standardized variables. Prove that the correlation between the predicted value $z_1' = rz_2$ and the actual value z_1 is equal to the correlation between X_1 and X_2.

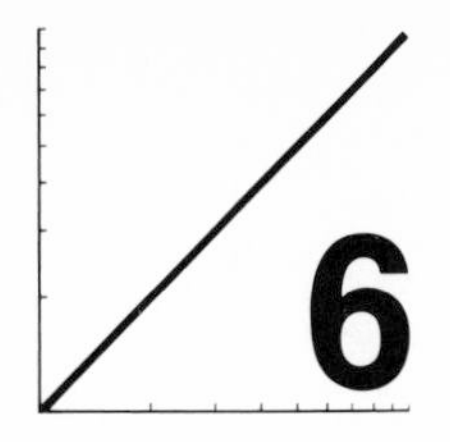

6 Factors Influencing the Magnitude of the Correlation Coefficient

6.1 The Shapes of the X and Y Distributions

One of the factors influencing the magnitude of a correlation coefficient is the shape or form of the distribution of the X and Y values. Only if the distributions of the X and Y values are symmetrical and have the same shape or form is it possible to pair the X and Y values in such a way as to result in a correlation coefficient of either -1.00 or 1.00. Figure 6.1 shows three different symmetrical frequency distributions of X and Y values. In Figure 6.1(*a*) both the X and Y distributions are symmetrical and both are bell shaped. It is possible, therefore, to pair the X and Y values in such a way as to result in a correlation coefficient of either -1.00 or 1.00. In Figure 6.1(*b*) both the X and Y distributions are symmetrical and both are U-shaped. Consequently, the X and Y values can be paired in such a way as to result in a correlation coefficient of either -1.00 or 1.00. In Figure 6.1(*c*) both the X and Y distributions are symmetrical and both are uniform or rectangular in shape. The X and Y values for these two distributions can also be paired so as to result in either $r = -1.00$ or $r = 1.00$.

In Figure 6.2(*a*) the frequency distributions of the X and Y values have the same shape or form, but the distributions are not symmetrical. Both distributions have the same degree of left skewness, that is, a tail to the left. In this example the X and Y values can be paired so as to result in a correlation coefficient equal to 1.00, but they cannot be paired in such a way as to result in a correlation coefficient of -1.00. Similarly, if the X and Y distributions have the same shape or form with the same degree of right skewness, that is, a tail to the right, then the values can be paired in such a way as to result in a correlation coefficient of 1.00 but not -1.00.

In Figure 6.2(*b*) the frequency distribution of the X values is left skewed and the frequency distribution of the Y values has the same degree of right skewness. In this example the X and Y values can be paired in such a way as to result in a correlation coefficient of -1.00, but they cannot be paired in such a way as to result in a value of 1.00.

6.2 Correlation Coefficients Based on Small Samples

A correlation coefficient based on a relatively small number of observations can be quite misleading. For example, if the population correlation coefficient is equal to zero and if random samples of $n = 5$ observations are drawn from the population, it can be expected that in 95 out of 100

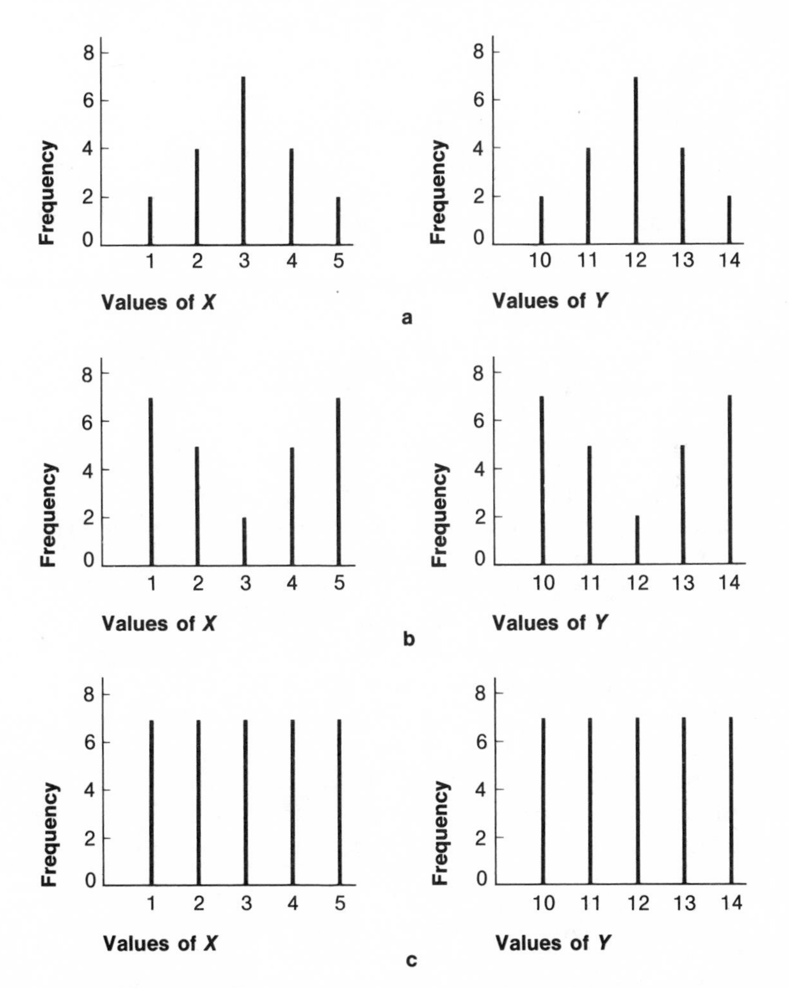

Figure 6.1 Three different symmetrical and identical frequency distributions for X and Y. The X and Y values for each frequency distribution can be paired so as to result in either $r = 1.00$ or $r = -1.00$.

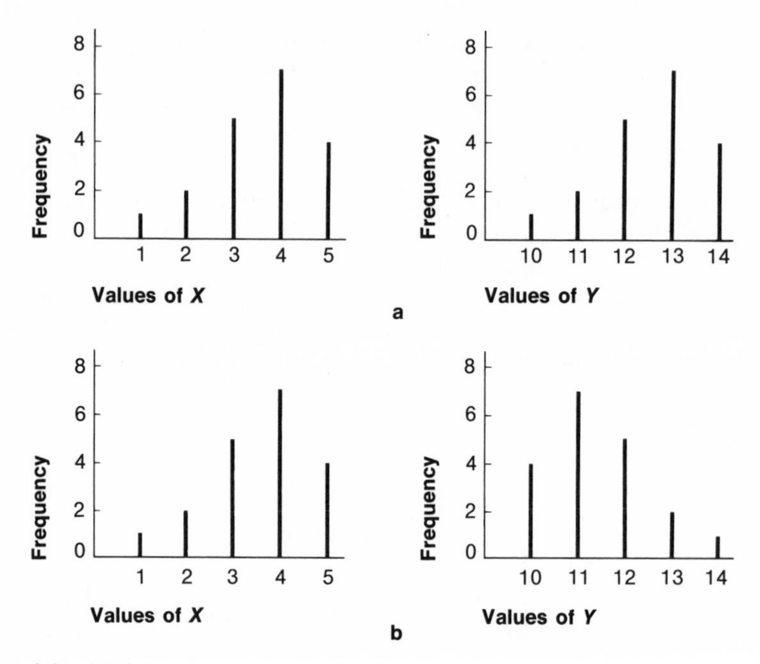

Figure 6.2 (*a*) A frequency distribution that has the same shape or form for both X and Y but that is not symmetrical. The X and Y values can be paired so as to result in $r = 1.00$ but not $r = -1.00$. (*b*) A frequency distribution for X that is left skewed and a frequency distribution for Y that has the same degree of right skewness. The X and Y values can be paired so as to result in $r = -1.00$ but not $r = 1.00$.

samples the correlation coefficient will fall within the range $-.88$ to .88. Thus, a relatively high correlation coefficient of .70 would not be at all unusual in a sample of $n = 5$ observations, even though the population correlation coefficient is equal to zero.

With small samples a single pair of (X, Y) values may contribute excessively to the value of the correlation coefficient. Consider, for example, the paired (X, Y) values shown in Figure 6.3. If the point with coordinates (8,8) is included, the correlation coefficient is equal to .74. If the point with coordinates (8,8) is omitted, the correlation coefficient is equal to zero.

With small samples, we should always be conscious of the fact that an excessively high negative or positive value of the correlation coefficient may simply be the result of an extreme pair of (X, Y) values.

6.3 Combining Several Different Samples

The magnitude of the correlation coefficient may also be influenced when a sample of observations really consists of two or more subsamples in which either the X means or the Y means, or both the X and Y means, differ from sample to sample. Before providing some examples of what may happen under these circumstances, we shall first review, briefly, the change in the two regression lines, Y on X and X on Y, as r increases from zero to one.

You will recall that when r is equal to zero, both b_X and b_Y are also equal to zero. Then the regression equation

$$Y' = \overline{Y} + b_Y(X - \overline{X})$$

results in a constant, $\overline{Y}$, for each value of X, and the "regression" line of Y on X is simply a horizontal line through the mean of the Y values. Similarly, if r is equal to zero, then the regression equation

$$X' = \overline{X} + b_X(Y - \overline{Y})$$

results in a constant, $\overline{X}$, for each value of Y, and the "regression" line of X on Y is simply a vertical line through the mean of the X distribution. These two lines are shown in Figure 6.4(*a*).

In Figure 6.4(*b*) we have $r = .20$. We have let $s_X = s_Y$ so that we also have $b_X = b_Y = r = .20$. Note that the two regression lines have moved slightly closer together. In Figure 6.4(*c*) we have $r = b_X = b_Y = .40$, and the two regression lines are closer together than they were when r was equal to .20. Figures 6.4(*d*) and 6.4(*e*) show the two regression lines when $r = b_X = b_Y = .60$ and when $r = b_X = b_Y = .80$, respectively. Note that the two regression lines move closer together as r increases. When $r = 1.00$, as shown in Figure 6.4(*f*), the two regression lines coincide. Similar changes in the two regression lines will occur with negative values of r, except that the two regression lines will then have negative slopes. When $r = -1.00$, the two regression lines will coincide.

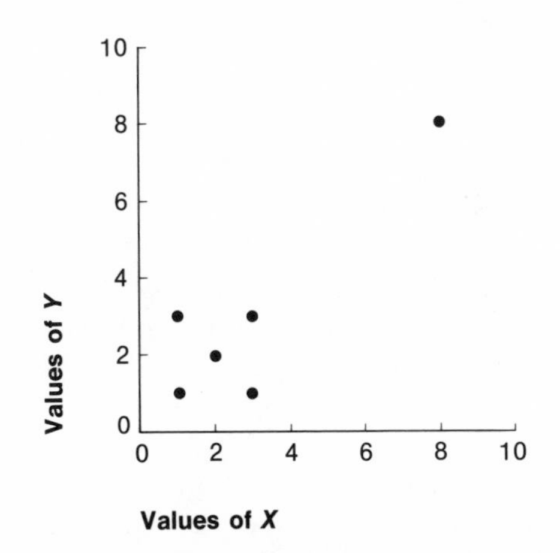

Figure 6.3 The correlation coefficient for the (X, Y) values is equal to .74. If the point with coordinates (8,8) is eliminated, the correlation coefficient is equal to zero.

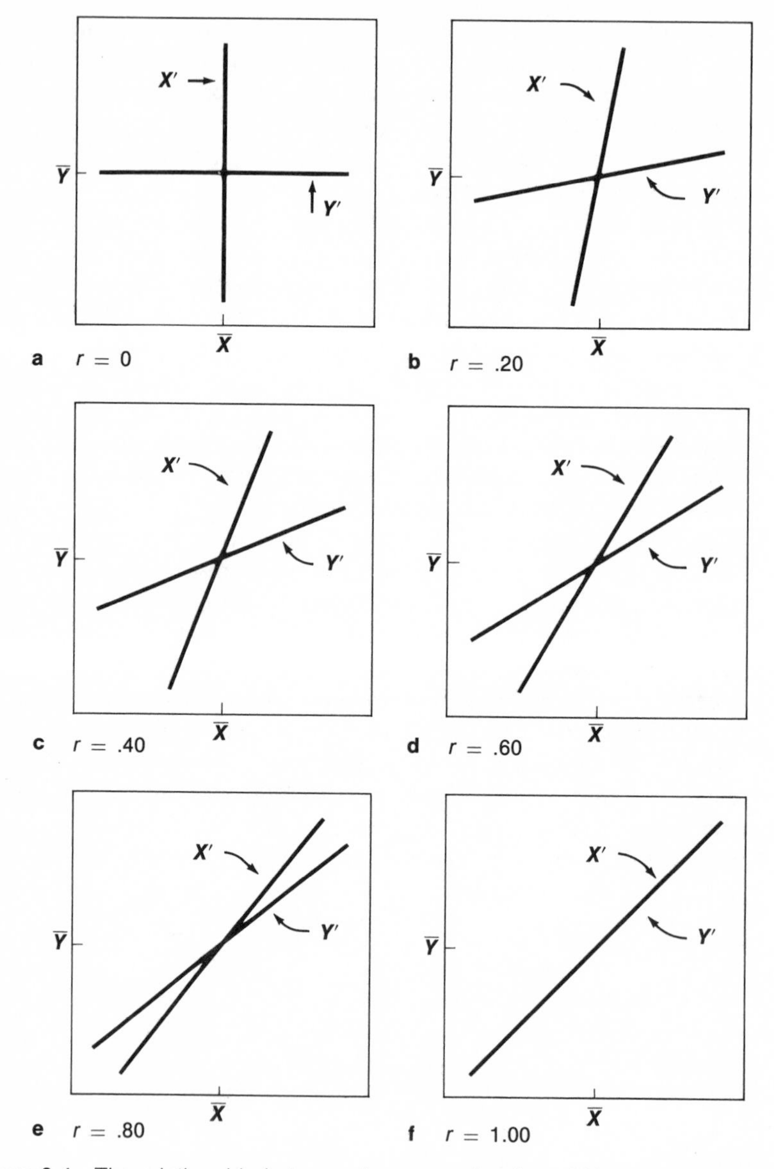

Figure 6.4 The relationship between the regression line of Y on X and the regression line of X on Y for various positive values of r. As r increases the two regression lines move closer together until, when $r = 1.00$, the two regression lines coincide.

Examine again Figure 6.4(*a*). With r equal to zero, the two regression lines are at right angles and can be enclosed in a circle. With r equal to .20, as shown in Figure 6.4(*b*), the two regression lines can be enclosed in a rather fat ellipse. Note that as r increases the ellipse enclosing the two regression lines becomes thinner and thinner. When r is equal to 1.00, the ellipse collapses into a single straight line.

In Figure 6.5(*a*) we have three samples. When each sample is considered separately, we have a fairly narrow ellipse, indicating that in each sample the correlation coefficient is relatively high. The ellipses slope upward from left to right, indicating that the correlation coefficients are positive. The X means for each sample are comparable but the Y means differ. If the three samples were combined we would have a relatively fat ellipse, and for the combined samples the correlation coefficient would be relatively low or close to zero.

The results just described may occur when a sample consists of both male and female subjects. If there are systematic differences in either the X means or the Y means for the male and female subjects, then the correlation coefficient obtained by combining the male and female subjects

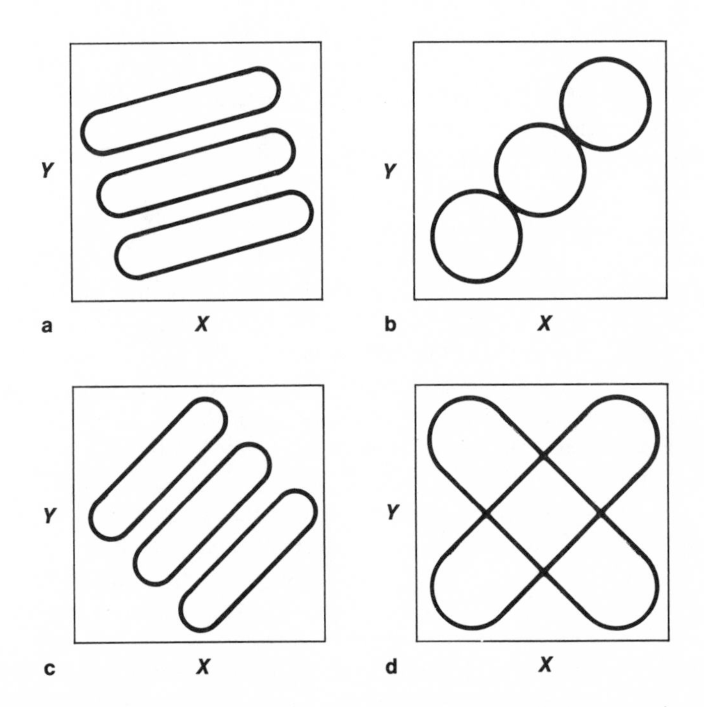

Figure 6.5 Diagrams illustrating how the correlation coefficient may be changed when different samples are combined. (*a*) In each sample the correlation coefficient is relatively high and positive. The correlation coefficient for the combined samples is close to zero. (*b*) In each sample the correlation coefficient is equal to zero. The correlation coefficient for the combined samples is moderately high and positive. (*c*) In each sample the correlation coefficient is high and positive. The correlation coefficient for the combined samples is low and negative. (*d*) In one sample the correlation coefficient is high and positive and in the other sample it is high and negative. The correlation coefficient for the combined samples is zero.

into a single sample may differ considerably from the correlation coefficients obtained for the male and female subjects separately.

Figure 6.5(*b*) illustrates the results obtained for three groups or samples. When each sample is considered separately, the correlation coefficient is zero, as indicated by the three circles. If the three samples were combined, we would have a moderate ellipse and the correlation coefficient for the combined samples would be positive and moderately high.

In Figure 6.5(*c*) the correlation coefficients for each of the three groups are relatively high and positive, as indicated by the three ellipses. In this example, the X and Y means for the three groups are negatively correlated. If the three samples were combined, the correlation coefficient would be low and negative.

Caution must always be exercised in interpreting a correlation coefficient based on a composite of several groups when the X means and/or the Y means are different for the subgroups making up the composite.

In Figure 6.5(*d*) the two samples have approximately the same X and Y means. But in one of the samples the correlation coefficient is high and positive and in the other sample the correlation coefficient is high and negative. If the two samples were combined, the correlation coefficient for the combined samples would be zero.

6.4 Restriction of Range

Figure 6.6 shows two variables with a moderately high positive correlation coefficient. Suppose now that the range of the X variable in our sample is restricted. For example, suppose that the sample consists only of those individuals with X values falling to the right of the vertical line in the figure. The restriction in the range of the X variable would result in a lower correlation than would be obtained if the range were not so restricted.

As a not unrealistic example, assume that in a sample of unselected seven-year-old children, IQ's and scores on a reading test are fairly highly positively correlated. A private school for gifted children, however, admits only those individuals aged seven years with IQ's equal to or greater than 130. This sample will have a restricted range of IQ's and is also likely to have a restricted range of scores on the reading test. For this sample the correlation coefficient between IQ's and scores on the reading test will be much lower than for the population of unselected seven-year-old children.

High school grades and first-year college grades of students in a given university may have a correlation of .60. If students who received low grades in high school are not admitted to the university, then the correlation coefficient of .60 is probably considerably lower than the correlation

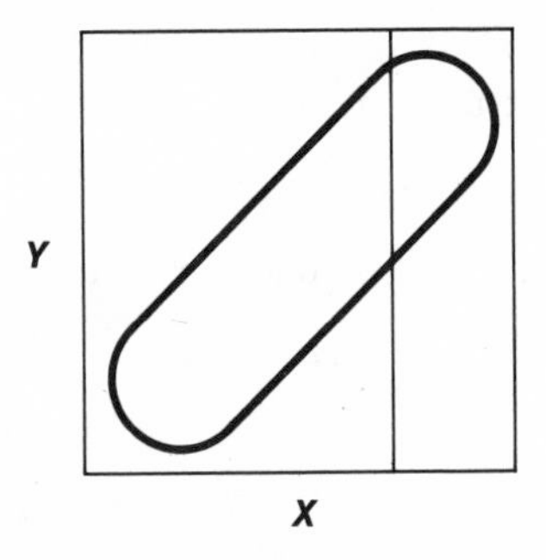

Figure 6.6 In the population of unselected observations, the correlation coefficient is moderately high and positive. If a sample has a restricted range on the X variable consisting of only those observations with values falling to the right of the vertical line, the correlation coefficient will be considerably lower.

coefficient that would have been obtained if all students were admitted regardless of their high school grades.

If a correlation coefficient in a population is not equal to zero, then the value of the correlation coefficient for a sample in which the range of cases is restricted by selecting only those cases with values of X or Y falling at one extreme will tend to be closer to zero than the population correlation coefficient for unselected cases whose range is not restricted.

6.5 Nonlinearity of Regression

When X and Y have a curvilinear relationship rather than a linear relationship, the correlation coefficient may be quite low and, in fact, not applicable. Thus if a low value of a correlation coefficient is obtained, it may be worthwhile to examine a plot of the Y values against the X values in order to determine whether the Y values are related to the X values, but not linearly related.

6.6 Correlation with a Third Variable: Partial Correlation

A relatively high or low value of a correlation coefficient between two variables, X_1 and X_2, may be obtained depending on the correlation coefficients of the two variables with a third variable X_3. In some instances we may be interested in determining the value of the correlation coefficient between X_1 and X_2 when X_3 is held constant. Suppose, for example, we obtain the regression equation.

$$x_1' = b_1 x_3$$

where $b_1 = \Sigma x_1 x_3 / \Sigma x_3^2$ and $x_1' = X_1' - \overline{X}_1$. Then for each individual we could obtain the residual error

$$x_1 - x_1' = x_1 - b_1 x_3$$

These residual errors will be uncorrelated with X_3. For example,

$$\begin{aligned}\Sigma(x_1 - x_1')x_3 &= \Sigma(x_1 - b_1 x_3)x_3 \\ &= \Sigma x_1 x_3 - b_1 \Sigma x_3^2 \\ &= 0\end{aligned}$$

and consequently $r_{(x_1 - x'_1)x_3}$ will be equal to zero. Similarly, we could obtain the regression equation

$$x_2' = b_2 x_3$$

where $b_2 = \Sigma x_2 x_3 / \Sigma x_3^2$ and $x_2' = X_2' - \overline{X}_2$. For each individual we could then obtain the residual error

$$x_2 - x_2' = x_2 - b_2 x_3$$

and these residuals could be shown in the same manner to be uncorrelated with X_3.

The correlation coefficient between the two sets of residuals represents the correlation between X_1 and X_2 with X_3 held constant. This correlation coefficient is called a *partial correlation coefficient*. The correlation coefficient between the two sets of residuals will be given by

$$r_{12.3} = \frac{r_{12} - r_{13} r_{23}}{\sqrt{1 - r_{13}^2}\,\sqrt{1 - r_{23}^2}} \tag{6.1}$$

where $r_{12.3}$ indicates the correlation between X_1 and X_2 with X_3 held constant.

For the proof of (6.1), without any loss of generality, we put all three variables in standard score form. Then the regression equation for predicting X_1 from X_3, in standard score form, will be

$$z_1' = r_{13} z_3$$

and the residual will be

$$z_1 - z_1' = z_1 - r_{13} z_3$$

Similarly, the regression equation for predicting X_2 from X_3, in standard score form, will be

$$z_2' = r_{23} z_3$$

and the residual will be

$$z_2 - z_2' = z_2 - r_{23} z_3$$

Then for the numerator of the correlation coefficient between the residuals, we have

$$\begin{aligned}\Sigma(z_1 - z_1')(z_2 - z_2') &= \Sigma(z_1 - r_{13}z_3)(z_2 - r_{23}z_3)\\ &= \Sigma z_1 z_2 - r_{23}\Sigma z_1 z_3 - r_{13}\Sigma z_2 z_3 + r_{13}r_{23}\Sigma z_3^2\end{aligned}$$

For the two terms in the denominator of the correlation coefficient, we have

$$\begin{aligned}\Sigma(z_1 - z_1')^2 &= \Sigma(z_1 - r_{13}z_3)^2\\ &= \Sigma z_1^2 - 2r_{13}\Sigma z_1 z_3 + r_{13}^2\Sigma z_3^2\end{aligned}$$

and

$$\begin{aligned}\Sigma(z_2 - z_2')^2 &= \Sigma(z_2 - r_{23}z_3)^2\\ &= \Sigma z_2^2 - 2r_{23}\Sigma z_2 z_3 + r_{23}^2\Sigma z_3^2\end{aligned}$$

Then the correlation coefficient between the residuals will be given by

$$r_{(z_1-z_1')(z_2-z_2')} = \frac{\Sigma z_1 z_2 - r_{23}\Sigma z_1 z_3 - r_{13}\Sigma z_2 z_3 + r_{13}r_{23}\Sigma z_3^2}{\sqrt{\Sigma z_1^2 - 2r_{13}\Sigma z_1 z_3 + r_{13}^2\Sigma z_3^2}\,\sqrt{\Sigma z_2^2 - 2r_{23}\Sigma z_2 z_3 + r_{23}^2\Sigma z_3^2}}$$

Dividing both the numerator and the denominator of this expression by $n - 1$, we have

$$\begin{aligned}r_{(z_1-z_1')(z_2-z_2')} &= \frac{r_{12} - r_{13}r_{23} - r_{13}r_{23} + r_{13}r_{23}}{\sqrt{1 - 2r_{13}^2 + r_{13}^2}\,\sqrt{1 - 2r_{23}^2 + r_{23}^2}}\\ &= \frac{r_{12} - r_{13}r_{23}}{\sqrt{1 - r_{13}^2}\,\sqrt{1 - r_{23}^2}}\end{aligned}$$

which is equal to (6.1).

The partial correlation coefficient $r_{12.3}$ will usually be smaller than the correlation coefficient r_{12}. For example, if we obtain the correlation coefficient between height and weight for children ranging in age from, say, 5 to 15 years, this correlation coefficient will be influenced by the fact that both height and weight are also positively correlated with age. Assume, for example, that X_1 represents height, X_2 represents weight, and X_3 represents age, and that the three correlation coefficients are $r_{12} = .90$, $r_{13} = .70$, and $r_{23} = .70$. Then the partial correlation coefficient between height and weight with age held constant will be

$$r_{12.3} = \frac{.90 - (.70)(.70)}{\sqrt{1 - .49}\,\sqrt{1 - .49}} = \frac{.41}{.51} = .80$$

As we have indicated, $r_{12.3}$ will, in general, tend to be smaller than r_{12}, but this is not necessarily always true. Suppose, for example, that with three variables either X_1 or X_2 has a zero correlation with X_3. Then, if $r_{12} = .80$, $r_{13} = .60$, and $r_{23} = 0$, we have

$$r_{12.3} = \frac{.80 - (.60)(.00)}{\sqrt{1 - .36}\ \sqrt{1}} = \frac{.80}{.80} = 1.00$$

In this interesting example, X_1 shares variance in common with X_3 but X_2 has no variance in common with X_3. Note that, in this instance,

$$z_1 - z_1' = z_1 - r_{13}z_3$$

removes the variance X_1 has in common with X_3 and that

$$z_2 - z_2' = z_2 - r_{23}z_3 = z_2$$

In essence, in this instance, the partial correlation coefficient removes the variance X_1 has in common with X_3 and the residuals are perfectly correlated with X_2. A variable X_3, which is correlated with X_1 or X_2 but has a zero correlation with the other variable, is commonly referred to as a *suppressor variable.* It suppresses the variance that either X_1 or X_2 has in common with X_3 but that is not part of the common variance of the other variable.

6.7 Random Errors of Measurement

Suppose that associated with each value of X_1 is a random error e_1 with mean equal to zero. Similarly, with each value of X_2 is associated a random error e_2 with mean equal to zero. We assume that the random errors are independent and also uncorrelated with the values of X_1 and X_2. In other words, if the values of X_1 and X_2 are put in deviation form, we assume that

$$\Sigma e_1 e_2 = \Sigma x_1 e_2 = \Sigma x_2 e_1 = \Sigma x_1 e_1 = \Sigma x_2 e_2 = 0 \tag{6.2}$$

Then the correlation coefficient between X_1 and X_2 will be

$$r_{12} = \frac{\Sigma(x_1 + e_1)(x_2 + e_2)}{\sqrt{\Sigma(x_1 + e_1)^2}\ \sqrt{\Sigma(x_2 + e_2)^2}}$$

$$= \frac{\Sigma x_1 x_2 + \Sigma x_1 e_2 + \Sigma x_2 e_1 + \Sigma e_1 e_2}{\sqrt{\Sigma x_1^2 + 2\Sigma x_1 e_1 + \Sigma e_1^2}\ \sqrt{\Sigma x_2^2 + 2\Sigma x_2 e_2 + \Sigma e_2^2}}$$

or, because of (6.2),

$$r_{12} = \frac{\Sigma x_1 x_2}{\sqrt{\Sigma x_1^2 + \Sigma e_1^2}\ \sqrt{\Sigma x_2^2 + \Sigma e_2^2}} \tag{6.3}$$

We see that the denominator of the correlation coefficient between X_1 and X_2 includes the two terms Σe_1^2 and Σe_2^2. Then r_{12} will take its maximum value (positive or negative) only if $\Sigma e_1^2 = \Sigma e_2^2 = 0$, and this can occur only if all values of $e_1 = e_2 = 0$. Thus, the presence of random errors in the measurements of X_1 and X_2 will obviously result in a lower value for the correlation coefficient r_{12}, than the value that would result in the absence of random errors in the measurements.

Measurements that are relatively free from random errors are commonly described as being reliable. There are obviously degrees of reliability, but the more reliable a set of measurements the smaller the error variance. If two psychological tests have relatively low reliabilities, that is, if each has a relatively large error variance, then the correlation between scores on the two tests will be considerably lower than the value that would be obtained if the two tests were more reliable.

Exercises

6.1. Under what conditions can X and Y values be paired in such a way as to result in either $r = 1.00$ or $r = -1.00$?

6.2. Assume that the correlation coefficient between X and Y for a sample of $n = 100$ males is equal to .60 and that, for a sample of $n = 100$ females, the correlation coefficient is equal to .70. The means of the X variable for both samples are approximately the same, but for the males $\overline{Y} = 60$ and for the females $\overline{Y} = 20$. If the two samples are combined, what would you predict about the correlation coefficient for the combined samples? Explain why.

6.3. The correlation coefficient in each of three independent samples is approximately equal to zero. The paired means $(\overline{X},\overline{Y})$ for the three samples are: (10,20), (20,30), and (30,40). What would you predict about the correlation coefficient for the combined samples? Explain why.

6.4. Assume that IQ's and scores on a reading comprehension test are relatively highly correlated, $r = .80$, for all third-grade children in a given school district. If the correlation coefficient between the same two variables is obtained for only those children with IQ's ranging from 90 to 110, what would you predict about this correlation coefficient? Explain why.

6.5. Suppose that the correlation coefficient, r_{13}, between X_1 and X_3 is equal to .70, and that the correlation coefficient, r_{23}, between X_2 and X_3 is equal to $-.70$. (a) Is it possible for the correlation coefficient, r_{12}, between X_1 and X_2 to be equal to .80? Explain why or why not. (b) Under the conditions described, what is the maximum possible positive value of r_{12}? (c) Under the conditions described, what is the maximum possible negative value of r_{12}?

6.6. Under what conditions might the partial correlation coefficient $r_{12.3}$ be higher than the correlation coefficient r_{12}?

6.7. Explain why random errors of measurement will tend to decrease the correlation coefficient between X_1 and X_2.

6.8. Suppose that $r_{12} = .80$, $r_{13} = .60$, and $r_{23} = 0$. Suppose also that we find

$$z_1' = r_{12}z_2 + r_{13}z_3$$

(a) Are the predicted values z_1' in standardized form? That is, is the mean equal to zero and the variance equal to one? (b) What is the value of the correlation coefficient between z_1 and z_1'?

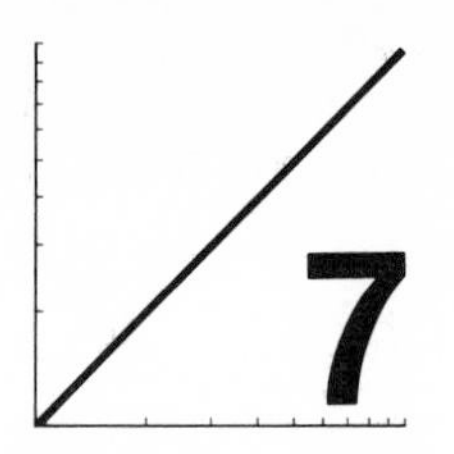

Special Cases of the Correlation Coefficient

7.1 Introduction

There are some variables for which observations can take only one of two possible values. Variables of this kind are called *dichotomous*, *binomial*, or *binary* variables. A common example is the response to items on tests where the only possible responses are True or False or where the response is classified as correct or incorrect. Other examples of dichotomous variables are subjects classified as male or female, employed or unemployed, and married or single.

In some instances we may have two such dichotomous variables and we wish to determine whether there is any relationship between the two variables. For example, if we have two items in a test we may wish to determine whether responses to the two items are independent or whether they are related. In other instances, we may have one variable that is dichotomous and another variable for which the observations can take any one of a number of different values. As an example, we may have subjects classified as male and female and for each subject we also have a score on a test. We wish to determine whether there is any relationship between the sex classification of the subjects and their scores on the test.

7.2 The Phi Coefficient

We consider first a measure of the degree to which two dichotomous variables are related. Suppose, for example, we have $n = 200$ subjects who have responded to two True-False items. We arbitrarily designate response to one of the items as the X variable and response to the other item as the Y variable. We also arbitrarily assign a value of 1 to a True response and a value of 0 to a False response for each item. We thus have two dichotomous variables, X and Y, for which the observations can take a value of either 1 or 0.

With two dichotomous variables of the kind described, the possible paired (X,Y) values are limited to the following cases: (1,0), (1,1), (0,0), and (0,1), as shown in Figure 7.1(a). Note that for Item 1, the X variable,

$$\Sigma X = n_1$$

is simply the number of individuals who have given the 1 response to this item, and that

$$\overline{X} = \frac{n_1}{n} = p_1 \tag{7.1}$$

or the proportion of individuals who have given the 1 response to the item.

We also have

$$\Sigma X^2 = n_1$$

and consequently

$$\begin{aligned} \Sigma(X - \overline{X})^2 &= \Sigma X^2 - \frac{(\Sigma X)^2}{n} \\ &= n_1^2 - \frac{n_1^2}{n} \end{aligned} \tag{7.2}$$

Similarly, for response to Item 2, the Y variable, we have

$$\Sigma Y = n_2, \qquad \Sigma Y^2 = n_2, \qquad \text{and } \overline{Y} = p_2$$

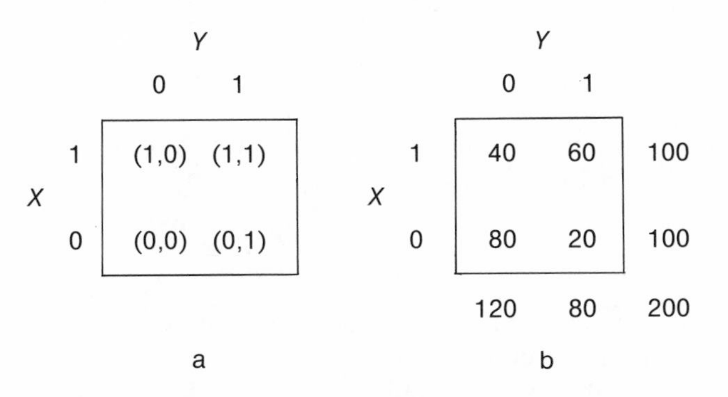

Figure 7.1 (a) The cell entries show the possible paired values of (X,Y) for two dichotomous variables. (b) The cell entries give the frequency of each of the paired (X,Y) values in a sample of $n = 200$.

where n_2 is the number of individuals who have given the 1 response to Item 2 and p_2 is the proportion of individuals who have given the 1 response to Item 2. Then, we also have

$$\Sigma(Y - \overline{Y})^2 = \Sigma Y^2 - \frac{(\Sigma Y)^2}{n}$$

$$= n_2 - \frac{n_2^2}{n}$$

To determine the sum of the products of the paired (X,Y) values, we note that for the pairs (1,0), (0,0), and (0,1), we have $XY = 0$, and that only the pairs (1,1) will contribute to ΣXY. We let n_{12} be the number of individuals with the values (1,1) and

$$\Sigma(X - \overline{X})(Y - \overline{Y}) = \Sigma XY - \frac{(\Sigma X)(\Sigma Y)}{n}$$

$$= n_{12} - \frac{n_1 n_2}{n} \tag{7.3}$$

We have

$$r = \frac{\Sigma XY - \frac{(\Sigma X)(\Sigma Y)}{n}}{\sqrt{\Sigma X^2 - \frac{(\Sigma X)^2}{n}}\sqrt{\Sigma Y^2 - \frac{(\Sigma Y)^2}{n}}} \tag{7.4}$$

and when there are two dichotomous variables, (7.4) becomes

$$r = \frac{n_{12} - \frac{n_1 n_2}{n}}{\sqrt{n_1 - \frac{n_1^2}{n}}\sqrt{n_2 - \frac{n_2^2}{n}}} \tag{7.5}$$

Dividing both the numerator and the denominator of (7.5) by n, we have

$$r = \frac{p_{12} - p_1 p_2}{\sqrt{p_1 - p_1^2}\sqrt{p_2 - p_2^2}} \tag{7.6}$$

where p_{12} is the proportion of individuals who have given the 1 response to both Item 1 and Item 2. If we let $q_1 = 1 - p_1$ and $q_2 = 1 - p_2$, then

$$r = \frac{p_{12} - p_1 p_2}{\sqrt{p_1 q_1}\sqrt{p_2 q_2}} \tag{7.7}$$

The cell entries in Figure 7.1(b) correspond to the number of individuals with each of the paired (X,Y) values shown in the cells of Figure 7.1(a). For this example, we have

$$p_1 = \frac{100}{200} = .50$$

$$p_2 = \frac{80}{200} = .40$$

$$p_{12} = \frac{60}{200} = .30$$

Then, substituting in (7.7), we obtain

$$r = \frac{.30 - (.50)(.40)}{\sqrt{(.50)(.50)}\ \sqrt{(.40)(.60)}} = .41$$

The value of r we have just obtained is the correlation coefficient between two variables for each of which the observations can take only values of 1 or 0. This correlation coefficient is frequently referred to as a *phi coefficient.*

The calculation of the correlation coefficient between two dichotomous variables can be considerably simplified and we now show how this is possible. Note, however, that (7.7) for two dichotomous variables is just a special case of one of the standard formulas for calculating the correlation coefficient.

Examine Figure 7.2. The letters a, b, c, and d in the cells of Figure 7.2(b) now correspond to the number of individuals with each of the paired (X,Y) values shown in the cells of Figure 7.2(a). We observe that we have the following identities:

$$\Sigma X = \Sigma X^2 = a + b$$
$$\Sigma Y = \Sigma Y^2 = b + d$$

and

$$\Sigma XY = b$$

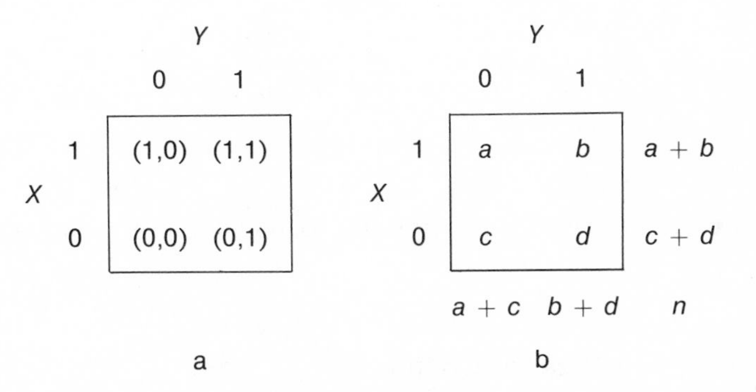

Figure 7.2 (a) The cell entries show the possible paired values of (X,Y) for two dichotomous variables. (b) The cell entries correspond to the frequency of each of the paired (X,Y) values in a sample of n observations.

We also have $n = a + b + c + d$. Substituting these identities in (7.4), we have

$$r = \frac{b - \dfrac{(a+b)(b+d)}{n}}{\sqrt{(a+b) - \dfrac{(a+b)^2}{n}}\sqrt{(b+d) - \dfrac{(b+d)^2}{n}}}$$

which can be shown to simplify to

$$r = \frac{bc - ad}{\sqrt{(a+c)(b+d)(a+b)(c+d)}} \tag{7.8}$$

If we substitute in the preceding expression the corresponding entries from Figure 7.1(b), we have

$$r = \frac{(60)(80) - (40)(20)}{\sqrt{(120)(80)(100)(100)}} = .41$$

as before.

7.3 Range of r for Dichotomous Variables

In our earlier discussion of the correlation coefficient, we said that r has a possible range from -1.00 to 1.00 only if the frequency distributions of both X and Y are symmetrical and have the same form. For dichotomous variables, the frequency distribution is symmetrical only if $p_1 = q_1 = .50$, that is, when the proportion of individuals with $X = 1$ is equal to the proportion with $X = 0$. If for both the X and Y variables we have

$$p_1 = p_2 = .50$$

then it is possible for r to be equal to either -1.00 or 1.00. On the other hand, if

$$p_1 = p_2 \neq .50$$

then it is possible for r to be equal to 1.00, but it is not possible for r to be equal to -1.00. If

$$p_1 = q_2 \neq .50$$

then it is possible for r to be equal to -1.00, but it is not possible for r to be equal to 1.00.

7.4 The Point Biserial Coefficient

Suppose that one variable (X) is a dichotomous variable for which the observations can take only the values of 1 or 0, but that for the other variable (Y) the observations can take any one of a number of different possible values. Given that the X variable has possible values of 1 and 0, we note that for all paired values $(0, Y)$ the product XY will be equal to zero and that for all paired values $(1, Y)$ the product XY will be equal to Y. Then

$$\Sigma XY = \Sigma Y_1$$

or the sum of the Y values for those observations with an X value equal to 1.

For the X variable, we have

$$\Sigma X = \Sigma X^2 = n_1$$

or the number of observations with the value of $X = 1$. Consequently,

$$\Sigma(X - \overline{X})^2 = n_1 - \frac{n_1^2}{n}$$

Then for the correlation coefficient between a binary variable (X) and a variable that can take a number of different possible values (Y), we have, by substitution in (7.4),

$$r = \frac{\Sigma Y_1 - \dfrac{n_1 \Sigma Y}{n}}{\sqrt{n_1 - \dfrac{n_1^2}{n}}\sqrt{\Sigma Y^2 - \dfrac{(\Sigma Y)^2}{n}}} \tag{7.9}$$

For the data of Table 7.1 we have $\Sigma Y_1 = 80$, $\Sigma Y = \Sigma Y_0 + \Sigma Y_1 = 30 + 80 = 110$, and

$$\Sigma Y^2 = (4)^2 + (3)^2 + \cdots + (8)^2 = 770$$

Substituting these values as well as $n_1 = 10$ and $n = n_0 + n_1 = 10 + 10 = 20$ in (7.4), we obtain

$$r = \frac{80 - \dfrac{(10)(110)}{20}}{\sqrt{10 - \dfrac{(10)^2}{20}}\sqrt{770 - \dfrac{(110)^2}{20}}} = \frac{25}{\sqrt{5}\sqrt{165}} = \frac{5}{\sqrt{33}} = .87$$

The correlation coefficient between a binary variable and a variable for which the observations can take any one of a number of different values is commonly referred to as the *point biserial coefficient.*

Note that the numerator of (7.9) can also be expressed as

$$\frac{n\Sigma Y_1 - n_1(\Sigma Y_0 + \Sigma Y_1)}{n} = \frac{1}{n}[\Sigma Y_1(n - n_1) - n_1 \Sigma Y_0]$$

$$= \frac{1}{n}(n_0 \Sigma Y_1 - n_1 \Sigma Y_0)$$

TABLE 7.1 Values of $(0, Y)$ and $(1, Y)$ for a dichotomous variable X and a variable Y that can take a number of different values for a sample of $n = 20$

	Values of Y	
	$(0, Y)$	$(1, Y)$
	4	8
	3	10
	3	10
	1	9
	2	6
	5	7
	5	7
	4	9
	1	6
	2	8
Σ	30	80

and that the first term in the denominator of (7.9) can be expressed as

$$\sqrt{\frac{nn_1 - n_1^2}{n}} = \sqrt{\frac{n_1 n_0}{n}}$$

Substituting these expressions in (7.9), we have

$$r = \frac{\frac{1}{n}(n_0 \Sigma Y_1 - n_1 \Sigma Y_0)}{\sqrt{\frac{n_1 n_0}{n}} \sqrt{\Sigma Y^2 - \frac{(\Sigma Y)^2}{n}}} \tag{7.10}$$

Multiplying the numerator and the denominator of (7.10) by $n/n_1 n_0$, we obtain

$$r = \frac{\overline{Y}_1 - \overline{Y}_0}{\sqrt{\frac{n}{n_1 n_0}} \sqrt{\Sigma Y^2 - \frac{(\Sigma Y)^2}{n}}} \tag{7.11}$$

Then for the data of Table 7.1 we also have

$$r = \frac{8 - 3}{\sqrt{\frac{20}{(10)(10)}} \sqrt{770 - \frac{(110)^2}{20}}} = \frac{5}{\sqrt{.2}\,\sqrt{165}} = \frac{5}{\sqrt{33}} = .87$$

as before.

As (7.11) shows, if $\overline{Y}_1$ is equal to $\overline{Y}_0$, then the point biserial coefficient will be equal to zero. If X is an ordered variable, then the point biserial coefficient will be positive if $\overline{Y}_1 > \overline{Y}_0$ and negative if $\overline{Y}_1 < \overline{Y}_0$. If X is an unordered variable, then the sign of the point biserial coefficient is irrelevant.

It can be shown, by using some tedious algebra, that the square of (7.9), (7.10), or (7.11) is equal to

$$r^2 = \frac{n_0(\bar{Y}_0 - \bar{Y})^2 + n_1(\bar{Y}_1 - \bar{Y})^2}{\Sigma(Y - \bar{Y})^2} \tag{7.12}$$

For the data of Table 7.1, for example, we have

$$r^2 = \frac{(10)(3 - 5.5)^2 + 10(8 - 5.5)^2}{165}$$

$$= \frac{125}{165}$$

and, consequently,

$$r = \sqrt{\frac{125}{165}} = \sqrt{\frac{25}{33}} = \frac{5}{\sqrt{33}} = .87$$

as before.

In the analysis of variance, $\Sigma(Y - \bar{Y})^2$, based on two or more sets of Y values, is commonly referred to as the total sum of squares, or SS_{tot}. In the simplest case of the analysis of variance, SS_{tot} is partitioned into two independent components, the treatment sum of squares, SS_T, and the pooled within treatment sum of squares, SS_W. When we have two groups or sets of Y values, as we do for the point biserial coefficient, the treatment sum of squares is equal to the numerator of (7.12), that is,

$$SS_T = n_0(\bar{Y}_0 - \bar{Y})^2 + n_1(\bar{Y}_1 - \bar{Y})^2 \tag{7.13}$$

Then we also have as a formula for the point biserial coefficient,

$$r = \sqrt{\frac{SS_T}{SS_{tot}}} \tag{7.14}$$

Because

$$SS_{tot} = SS_T + SS_W \tag{7.15}$$

the pooled within treatment sum of squares can be obtained by subtraction. For the data of Table 7.1, for example, we have

$$SS_W = SS_{tot} - SS_T = 165 - 125 = 40$$

We can, of course, calculate SS_W. In our example it is equal to the sum of

$$\Sigma y_0^2 = \Sigma Y_0^2 - \frac{(\Sigma Y_0)^2}{n_0}$$

$$= 110 - \frac{(30)^2}{10}$$

$$= 20$$

and

$$\Sigma y_1^2 = \Sigma Y_1^2 - \frac{(\Sigma Y_1)^2}{n_1}$$

$$= 660 - \frac{(80)^2}{10}$$

$$= 20$$

or

$$SS_W = \Sigma y_0^2 + \Sigma y_1^2 = 20 + 20 = 40$$

It is obvious from (7.14) that the point biserial coefficient can be equal to 1.00 only if SS_T is equal to SS_{tot} and, in this case, SS_W would have to be equal to zero. But the only way in which SS_W can be equal to zero is if the n_0 values of $X = 0$ all have exactly the same Y_0 value and if the n_1 values of $X = 1$ also all have the same Y_1 value. In this instance, the point biserial coefficient would, in essence, be the same as the correlation coefficient between two binary variables for which r is equal to either -1.00 or 1.00.

In general, the Y values for the two categories of X will not have zero variance and, consequently, the point biserial coefficient will not be equal to 1.00. In the example shown in Table 7.1, we have maximized the value of the point biserial coefficient for the given values of Y. There is, in other words, no way in which these same Y values could be rearranged to obtain a value for the point biserial coefficient higher than .87.

7.5 The Rank Order Correlation Coefficient

Another special case of the correlation coefficient is when both the X and Y variables consist of a set of ranks. Suppose, for example, that two judges have ranked the same set of n objects according to some property of interest. We are interested in determining whether the ranks assigned to the objects by one judge are related to or show any agreement with the ranks assigned to the same objects by another judge.

If the values of a variable X consist of the ranks from 1 to n, then it can be shown that

$$\Sigma X = \frac{n(n + 1)}{2} \tag{7.16}$$

and that

$$\Sigma(X - \overline{X})^2 = \frac{n^3 - n}{12} \tag{7.17}$$

For the data of Table 7.2, we have

$$\Sigma X = \Sigma Y = \frac{10(10 + 1)}{2} = 55$$

TABLE 7.2 Rank values of (X,Y) in a sample of $n = 10$

	X	Y	XY	$D = X - Y$	D^2
	10	8	80	2	4
	9	10	90	−1	1
	8	9	72	−1	1
	7	7	49	0	0
	6	4	24	2	4
	5	6	30	−1	1
	4	5	20	−1	1
	3	3	9	0	0
	2	1	2	1	1
	1	2	2	−1	1
Σ	55	55	378	0	14

and

$$\Sigma(X - \overline{X})^2 = \Sigma(Y - \overline{Y})^2 = \frac{10^3 - 10}{12} = 82.5$$

We also have

$$\Sigma XY = 378$$

Substituting these values in (7.4), we obtain

$$r = \frac{378 - \dfrac{(55)(55)}{10}}{\sqrt{82.5}\,\sqrt{82.5}} = \frac{75.5}{82.5} = .915$$

The correlation coefficient between two sets of ranks is commonly referred to as the *rank order correlation coefficient*. The calculation of the rank order correlation coefficient can be considerably simplified. For example, if for *any* pair of (X,Y) values we let

$$D = X - Y$$

then

$$\Sigma D = \Sigma X - \Sigma Y$$

and

$$\overline{D} = \overline{X} - \overline{Y}$$

The sum of the squared deviations of the D values from the mean of the D values will be

$$\begin{aligned}\Sigma(D - \overline{D})^2 &= \Sigma[(X - \overline{X}) - (Y - \overline{Y})]^2\\ &= \Sigma x^2 + \Sigma y^2 - 2\Sigma xy\\ &= \Sigma x^2 + \Sigma y^2 - 2\Sigma xy\frac{\sqrt{(\Sigma x^2)(\Sigma y^2)}}{\sqrt{(\Sigma x^2)(\Sigma y^2)}}\\ &= \Sigma x^2 + \Sigma y^2 - 2r\sqrt{(\Sigma x^2)(\Sigma y^2)}\end{aligned}$$

and

$$r = \frac{\Sigma x^2 + \Sigma y^2 - \Sigma(D - \overline{D})^2}{2\sqrt{(\Sigma x^2)(\Sigma y^2)}} \tag{7.18}$$

Substituting in the preceding expression the corresponding values for ranks, we obtain

$$r = \frac{\dfrac{n^3 - n}{12} + \dfrac{n^3 - n}{12} - \Sigma(D - \overline{D})^2}{2\sqrt{\left(\dfrac{n^3 - n}{12}\right)\left(\dfrac{n^3 - n}{12}\right)}}$$

or

$$r = 1 - \frac{6\Sigma(D - \overline{D})^2}{n^3 - n} \tag{7.19}$$

But if X and Y are ranks, then $\overline{X} = \overline{Y}$ and

$$\overline{D} = \overline{X} - \overline{Y} = 0$$

and, consequently,

$$r = 1 - \frac{6\Sigma D^2}{n^3 - n} \tag{7.20}$$

For the data of Table 7.2 we have

$$r = 1 - \frac{(6)(14)}{10^3 - 10} = .915$$

which is the same value we obtained before.

7.6 The Variance and Standard Deviation of the Difference between Two Independent Variables

We have shown that for *any* pair of (X,Y) values, if $D = X - Y$, then

$$\Sigma(D - \overline{D})^2 = \Sigma x^2 + \Sigma y^2 - 2\Sigma xy$$

Dividing both sides of this expression by $n - 1$, we have the variance of the differences or

$$s_D^2 = s_{X-Y}^2 = s_X^2 + s_Y^2 - 2c_{XY}$$

It can be shown that if X and Y are independent variables, then c_{XY} will be equal to zero. Consequently, if X and Y are independent variables, then the variance of the difference will be

$$s_{X-Y}^2 = s_X^2 + s_Y^2$$

and the standard deviation will be

$$s_{X-Y} = \sqrt{s_X^2 + s_Y^2} \tag{7.21}$$

We shall use variations of (7.21) in later chapters, where we will be interested in testing the significance of the difference between two correlation coefficients, two regression coefficients, or two means based on two independent random samples.

Exercises

7.1. We have two dichotomous variables, X_1 and X_2. The numbers of observations in a sample of $n = 200$ with the response patterns (1,0), (1,1), (0,0) and (0,1) are 45, 45, 80, and 30, respectively. Find the value of the correlation coefficient between X_1 and X_2.

7.2. We know the scores of eight males and eight females on a test Y. We arbitrarily assign a value of $X = 0$ to the males and $X = 1$ to the females. The scores of the subjects on the test (Y) are as follows:

Males (X_0)	Females (X_1)
7	10
6	9
5	8
4	7
4	7
3	6
2	5
1	4

Find the value of the correlation coefficient between X and Y.

7.3. A group of male students and a group of female students rated eight personality traits in terms of their desirability in a marital partner. The average ratings for each trait for each group of subjects were then translated into ranks. The ranks assigned to the traits by the males and females are as follows:

Trait	Male	Female
A	1	7
B	2	3
C	3	1
D	4	5
E	5	8
F	6	4
G	7	6
H	8	2

Calculate the correlation coefficient between the two sets of ranks.

7.4. Two judges tasted and ranked each of nine domestic brands of red wine in terms of overall merit. The ranks assigned to the brands by each judge are as follows:

Brand	Judge 1	Judge 2
A	1	6
B	3	5
C	2	2
D	6	1
E	4	8
F	5	3
G	8	7
H	9	4
I	7	9

Calculate the correlation coefficient between the two sets of ranks.

7.5. A study of 100 women who thought their marriage was successful and 100 women who thought their marriage was unsuccessful revealed a differential in response to the question: Did you have a happy childhood? For these two dichotomous variables, the following frequencies were obtained:

	Marital status	
Childhood status	Unsuccessful	Successful
Happy	40	70
Unhappy	60	30

Find the value of the correlation coefficient between the two dichotomous variables.

7.6. Under what conditions can the phi coefficient be equal to either 1.00 or −1.00?

7.7. Under what conditions can the phi coefficient be equal to 1.00 but not −1.00?

7.8. Under what conditions can the phi coefficient be equal to −1.00 but not 1.00?

7.9. If $\overline{Y}_0 = \overline{Y}_1$, then what do we know about the value of the point biserial coefficient?

7.10. Under what conditions can the point biserial coefficient be equal to 1.00?

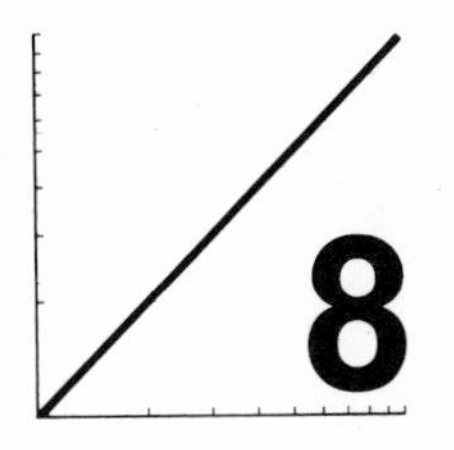

8 Tests of Significance for Correlation Coefficients

8.1 Introduction

The calculation of the correlation coefficient r for a sample of n observations requires no assumptions about the distributions of either the X or Y variables, although the nature of these distributions may place limitations on the magnitude of the correlation coefficient in the manner in which we have indicated previously. If, however, we are interested in using the sample value of r to infer something about the corresponding population parameter ρ, then we do need to be concerned about distribution assumptions.

In a typical correlation problem, we assume that we have a random sample of n paired (X,Y) values. Note that the values of X are not preselected as in a typical regression problem. On the basis of the obtained sample correlation coefficient r, we wish to test a null hypothesis regarding the population value ρ. In order to use the available tests of significance, we must make certain assumptions regarding the distribution of both the X and the Y values. First of all, it is assumed that both the X and Y variables are normally distributed in the population with corresponding means μ_X and μ_Y and variances σ_X^2 and σ_Y^2. It is not necessary that $\mu_X = \mu_Y$ or that $\sigma_X^2 = \sigma_Y^2$. For each possible value of X there is a corresponding population of normally distributed Y values with mean $\mu_{Y.X}$ and variance $\sigma_{Y.X}^2$. The variance $\sigma_{Y.X}^2$ is assumed to be the same for each value of X. Similarly, for each possible value of Y there is a corresponding population of X values with mean $\mu_{X.Y}$ and with variance $\sigma_{X.Y}^2$, which is assumed to be the same for each value of Y. When these assumptions are met, the joint distribution of the paired (X,Y) values is said to be a *bivariate normal distribution.*

In general, tests of significance of correlation coefficients are based on the assumption that X and Y have a bivariate normal distribution.[1] With a bivariate normal distribution, the only possible relationship between X and Y is a linear relationship. It does not follow, however, that X and Y must be linearly related; that is, it is possible for ρ to be equal to zero, even though the population distribution is bivariate normal.

8.2 Tests of Significance

Ordinarily, in testing a null hypothesis for significance we decide in advance on some small probability, called the *significance level* of the test and represented by α, as a standard for deciding whether to reject the null hypothesis. Frequently used standards are $\alpha = .05$ and $\alpha = .01$, but other values of α might also be chosen. If the result of the test of significance is such that the probability of the outcome is equal to or less than α, then the null hypothesis is rejected. This simply means that the probability of the outcome is sufficiently small that we choose to regard the null hypothesis as improbable or false.

A Type I error occurs when a null hypothesis is, in fact, true, but the test of significance results in a decision to reject it.[2] Our protection against making a Type I error is equal to $1 - \alpha$. For example, if we set $\alpha = .01$, this is the probability that we will make a Type I error, given that the null hypothesis is true; and our protection against making a Type I error, given that the null hypothesis is true, is equal to $1 - \alpha = .99$.

Tests of significance are commonly made by transforming a statistic or statistics based on a sample into another statistic for which the probability distribution is known, given that a null hypothesis is true. In this chapter, we make use of three different probability distributions: the standard normal distribution Z, the t distribution, and the χ^2 distribution. In Chapter 9 we make use of another probability distribution, the F distribution.

Each of these distributions is represented by a curve such that the area under the curve is equal to one. Tables in the Appendix give the proportion of the total area in the right tail of the curve when ordinates are erected at various tabled values of Z, t, χ^2, and F. These areas correspond to the probabilities of obtaining Z, t, χ^2, or F equal to or greater than the tabled value, when a null hypothesis is true, and can be used to test various null hypotheses.

---

[1]An exception, as we shall show in Chapter 10, is the test of the null hypothesis that $\rho = 0$.

[2]We should also be concerned about Type II errors. A Type II error occurs when a null hypothesis is false but the test of significance fails to reject it. Other things being equal, the probability of a Type II error decreases as the sample size increases.

There is only one standard normal distribution and, consequently, the table of the standard normal distribution is entered only in terms of Z. The t, χ^2, and F distributions depend on the number of degrees of freedom involved. Consequently, in using the t, χ^2, and F tables we need to know not only the observed value of t, χ^2, or F, but also the number of degrees of freedom associated with the observed value.

8.3 Sampling Distribution of the Correlation Coefficient

Suppose that there exists a population of paired (X,Y) values such that for this population the correlation coefficient is equal to ρ. If random samples of n are drawn from this population, each of the n observations will consist of an ordered pair of (X,Y) values, and for each random sample it will be possible to calculate the sample correlation coefficient r. If ρ is close to zero, then the sampling distribution of the r values will be approximately normal in form, provided n is not too small. However, if ρ is not close to zero and if n is small, then the sampling distribution of r will not be normal in form but will instead be skewed. For example, if $\rho = 0.80$ and $n = 8$, the sample values will tend to cluster around 0.80, but there will be a tail to the left; that is, the distribution will be left skewed.

8.4 Test of the Null Hypothesis that $\rho = 0$

Table 8.1 shows the X and Y values for two independent random samples. For the first sample we have $n = 15$ observations and for the second sample we have $n = 10$ observations. For the first sample, we have

$$r = \frac{564 - (63)(110)/15}{\sqrt{365 - (63)^2/15}\sqrt{934 - (110)^2/15}} = \frac{102}{\sqrt{100.40}\sqrt{127.33}} = .902$$

To test the null hypothesis that ρ, the population correlation coefficient, is equal to zero, we calculate

$$t = \frac{r}{\sqrt{1 - r^2}}\sqrt{n - 2} \tag{8.1}$$

Then t, as defined by (8.1), will be distributed in accordance with the tabled values of t with degrees of freedom equal to $n - 2$, when the null hypothesis is true. For example, with $15 - 2 = 13$ d.f., we find from the table of t, Table IV in the Appendix, that the probability of obtaining $t \geq 3.012$ is .005, if it is true that $\rho = 0$. Because the distribution of t is symmetrical, when the null hypothesis is true, the probability of obtaining $t \leq -3.012$ is also .005. Then the probability of obtaining $t \geq 3.012$ or $t \leq -3.012$ is equal to $.005 + .005 = .01$, if $\rho = 0$.

TABLE 8.1 Values of (X,Y) for two independent random samples of $n = 15$ and $n = 10$

	Sample 1: $n = 15$				
	X	Y	X^2	Y^2	XY
	6	10	36	100	60
	1	6	1	36	6
	1	3	1	9	3
	1	4	1	16	4
	6	9	36	81	54
	7	10	49	100	70
	2	3	4	9	6
	4	8	16	64	32
	8	11	64	121	88
	8	11	64	121	88
	1	6	1	36	6
	5	10	25	100	50
	7	10	49	100	70
	3	5	9	25	15
	3	4	9	16	12
Σ	63	110	365	934	564

	Sample 2: $n = 10$				
	X	Y	X^2	Y^2	XY
	3	2	9	4	6
	1	4	1	16	4
	1	2	1	4	2
	7	7	49	49	49
	6	6	36	36	36
	3	2	9	4	6
	1	4	1	16	4
	5	6	25	36	30
	8	10	64	100	80
	7	8	49	64	56
Σ	42	51	244	329	273

Let us assume that we have chosen $\alpha = .01$. Because we are interested in the possibility that ρ may be either positive or negative, if it is not equal to zero, we will reject the null hypothesis if either $t \geq 3.012$ or $t \leq -3.012$. For our example, we have

$$t = \frac{.902}{\sqrt{1 - (.902)^2}} \sqrt{15 - 2} = 7.53$$

Because our obtained $t = 7.53$ exceeds the tabled value 3.012, we reject the null hypothesis.[3]

[3]From the table of t, Table IV in the Appendix, we find that the probability of obtaining t equal to or greater than 7.53, when the null hypothesis is true, is approximately .0000025. Consequently, the probability of obtaining $t \geq 7.53$ or $t \leq -7.53$ is approximately .00005.

8.5 Table of Significant Values of r

It is possible to substitute in (8.1) the various values of n and the tabled values of t and to solve for the values of r that would be required at various levels of significance. This has been done and the resulting values of r to three decimal places are given in Table V in the Appendix. The values of r given in Table V are those that would be regarded as significant with probabilities given by the column headings if one-sided tests are made using the right tail of the t distribution, that is, for positive values of r. Because the distribution of r is symmetrical when $\rho = 0$, the table can also be used with negative values of r. For example, with $n - 2 = 13$ d.f., we find that the probability of obtaining $r \geq .641$ is .005 when $\rho = 0$. This is also the probability of obtaining $r \leq -.641$ when $\rho = 0$. For a two-sided test of the null hypothesis, the probability of obtaining either $r \geq .641$ or $r \leq -.641$ is .01.

It is evident from Table V that a relatively large observed value of r may not be significant when r is based on a small number of observations. On the other hand, as n increases, relatively small values of r will result in the rejection of the null hypothesis. For example, if a correlation coefficient is based on $n = 1000$ observations, then values of r equal to or greater than .06 or equal to or less than $-.06$ have a probability of approximately .05, when the null hypothesis is true.

8.6 The z_r Transformation for r

Any value of r may be transformed to a new variable z_r, defined as

$$z_r = \tfrac{1}{2}[\log_e(1 + r) - \log_e(1 - r)] \tag{8.2}$$

where r is the observed value of the correlation coefficient.[4] In order to make the z_r values of (8.2) available independently of a table of natural logarithms, values of r were substituted in (8.2) and the corresponding values of z_r were obtained. These z_r values are given in Table VI in the Appendix. In our example, we have $r = .902$, and from Table VI we find that the corresponding value of z_r is approximately 1.472. The z_r distribution is symmetrical about zero, and if a correlation coefficient has a minus sign, then so also will the corresponding value of z_r. For example, if $r = -.902$, then $z_r = -1.472$.

Fisher[5] has shown that the distribution of z_r is approximately normal in form and that for all practical purposes the distribution is independent

---

[4]The notation z_r is used to indicate a transformation for r, but z_r should not be confused with a standardized variable, which has a mean equal to zero and a variance equal to one and for which we have previously used the symbol z.

[5]Fisher, R. A. On the "probable error" of a coefficient of correlation deduced from a small sample. *Metron*, 1921, *1*, Part 4, 1–32.

of the population value ρ. This means that the distribution of z_r remains approximately normal in form even when samples are drawn from populations in which ρ is large. Furthermore, the standard error of z_r is related in a very simple way to n, the sample size, and is given by

$$\sigma_{z_r} = \frac{1}{\sqrt{n-3}} \tag{8.3}$$

where n is the number of observations on which r is based. For the first sample in Table 8.1 we have $n = 15$ observations and

$$\sigma_{z_r} = \frac{1}{\sqrt{15-3}} = .29$$

If z_r is approximately normally distributed, then

$$Z = \frac{z_r - z_\rho}{\sigma_{z_r}} \tag{8.4}$$

will have a distribution that is approximately that of a standard normal variable with $\mu = 0$ and $\sigma = 1$ and can be evaluated in terms of the table of the standard normal distribution, Table II in the Appendix.[6] In (8.4), z_r is the value of z corresponding to the observed value of r and z_ρ is the value corresponding to ρ.

Suppose we want to make a two-sided test of the null hypothesis that $\rho = .60$, with $\alpha = .01$. From Table VI we find that $z_\rho = .693$. From the table of the standard normal distribution, Table II, we find that the probability of obtaining $Z \geq 2.58$, when the null hypothesis is true, is .005, and this is also the probability of obtaining $Z \leq -2.58$. Thus, with a two-sided test, we will reject the null hypothesis if we obtain Z equal to or greater than 2.58 or Z equal to or less than -2.58. Substituting the values $z_r = 1.472$, $z_\rho = .693$, and $\sigma_{z_r} = .29$ in (8.4), we have

$$Z = \frac{1.472 - .693}{.29} = 2.686$$

and because Z is greater than 2.58, the null hypothesis is rejected. We conclude that it is improbable that the population correlation is as low as .60.

8.7 Establishing a Confidence Interval for ρ

Rather than testing various null hypotheses regarding the value of ρ, suppose that we set up the following inequality:

---

[6]For a standardized variable that is also normally distributed we use the symbol Z rather than z.

$$-2.58 < \frac{z_r - z_\rho}{\sigma_{z_r}} < 2.58$$

If $z_r = 1.472$ and $\sigma_{z_r} = .29$, we have

$$-2.58 < \frac{1.472 - z_\rho}{.29} < 2.58$$

or

$$1.472 + (.29)(2.58) > z_\rho > 1.472 - (.29)(2.58)$$
$$2.220 > z_\rho > .724$$

From the table of z_r we find that the two r's corresponding to $z_r = 2.220$ and $z_r = .724$ are approximately .975 and .620, respectively.

The interval, .620 to .975, that we have just established is called a 99 percent *confidence interval* for ρ. The value .620 is called the *lower confidence limit* and the value .975 is called the *upper confidence limit.* Confidence limits are statistics and, like all statistics, may be expected to vary from sample to sample. If we have a new and independent sample of $n = 15$ from the same population as the present one, we will not necessarily find that r for this sample is equal to .902 and, consequently, the lower and upper 99 percent confidence limits will not be the same as those obtained with the present sample. We can, however, make the statement that we are 99 percent confident that a 99 percent confidence interval will contain ρ. The basis for this statement is that if we have an indefinitely large number of samples, we can expect that in 99 out of 100 samples, the 99 percent confidence interval will contain ρ and that in 1 out of 100 samples, ρ will not be contained within the interval. We, of course, have no way of knowing whether the obtained interval is one of the 99 in 100 that contain ρ or whether it is the 1 in 100 that does not. However, if we always infer that ρ falls within a 99 percent confidence interval, we can expect that, in the long run, 99 percent of our inferences will be correct and only 1 percent will be incorrect.

We note that the confidence limits, .620 and .975, are not equally distant from the observed value of $r = .902$. The lower confidence limit deviates .282 from the observed value of r and the upper confidence limit deviates .073 from r. However, as n increases, the lower and upper confidence limits on the r scale will become more symmetrical about the observed value of r. For example, if $r = .902$ were based on a sample of $n = 403$ observations, then $\sigma_{z_r} = 1/\sqrt{403 - 3} = .05$. If we now find a 99 percent confidence interval for ρ, we will observe that the lower limit on the r scale is approximately .87 and the upper limit is approximately .92. These two values, based on a sample of $n = 403$ observations, are more symmetrical about the observed value of $r = .902$ than are the two values based on a sample of $n = 15$ observations. What this indicates, of course, is that as n increases, the skewness of the sampling distribution of r decreases.

8.8 Test of Significance of the Difference between r_1 and r_2

One of the advantages of the z_r transformation for r is that it also permits us to test the significance of the difference between two values of r obtained from two independent samples. To illustrate the test of significance, we use the data given in Table 8.1 for two samples. We have already found that for the first sample of $n = 15$ observations, $r_1 = .902$. For the second sample, we have

$$r_2 = \frac{273 - (42)(51)/10}{\sqrt{244 - (42)^2/10}\,\sqrt{329 - (51)^2/10}} = \frac{58.8}{\sqrt{(67.6)(68.9)}} = .862$$

The two values $r_1 = .902$ and $r_2 = .862$ are not equal. Is it reasonable to believe that the two samples are from a common population so that $\rho_1 = \rho_2$? If we reject this null hypothesis, we will conclude that $\rho_1 \neq \rho_2$; in other words, that the difference between $r_1 = .902$ and $r_2 = .862$ is sufficiently large that we do not believe they are both estimates of the same population value ρ.

To make the test of significance, we transform both r_1 and r_2 into z_r values. The standard error of the difference between two independent values of z_r will be given by the usual formula for the standard error of the difference between two independent variables[7] or, in the case of two z_r values, z_1 and z_2,

$$\begin{aligned} \sigma_{z_1 - z_2} &= \sqrt{\sigma_{z_1}^2 + \sigma_{z_2}^2} \\ &= \sqrt{\frac{1}{n_1 - 3} + \frac{1}{n_2 - 3}} \end{aligned} \tag{8.5}$$

In our example, we have $n_1 = 15$ and $n_2 = 10$, and therefore

$$\sigma_{z_1 - z_2} = \sqrt{\frac{1}{15 - 3} + \frac{1}{10 - 3}} = .476$$

Then, the difference between z_1 and z_2 divided by the standard error of the difference results in

$$Z = \frac{z_1 - z_2}{\sigma_{z_1 - z_2}} \tag{8.6}$$

If the null hypothesis $\rho_1 = \rho_2$ is true, then Z will have a distribution that is approximately that of a standard normal variable with $\mu = 0$ and $\sigma = 1$ and can be evaluated in terms of the table of the standard normal distribution, Table II in the Appendix.

---

[7]See Section 7.6.

In our example, we have $r_1 = .902$ with $z_1 = 1.472$ and $r_2 = .862$ with $z_2 = 1.293$. Then, substituting these two values of z_r and $\sigma_{z_1-z_2} = .476$ in (8.6), we have

$$Z = \frac{1.472 - 1.293}{.476} = .376$$

For a two-sided test, using the table of the standard normal distribution, we find that the probability of $Z \geq .376$ or $Z \leq -.376$ is about .71, when the null hypothesis $\rho_1 = \rho_2$ is true. We may regard the null hypothesis as tenable and conclude that the difference between the two correlation coefficients is not sufficiently great to cause us to believe that they are not both estimates of the same population value ρ.

Because the two values of r can be considered estimates of the same population value ρ, we calculate

$$\bar{z}_r = \frac{(n_1 - 3)z_1 + (n_2 - 3)z_2}{(n_1 - 3) + (n_2 - 3)} \tag{8.7}$$

which is the weighted average value of z_r. The weighted average in the present problem is

$$\bar{z}_r = \frac{(15 - 3)(1.472) + (10 - 3)(1.293)}{(15 - 3) + (10 - 3)} = \frac{26.715}{19} = 1.406$$

From Table VI we find that the corresponding value of r is approximately .885. We may regard this value as an estimate of the common population value ρ based on the data from the two samples.

8.9 The χ^2 Test for the Difference between r_1 and r_2

We now consider an equivalent method for testing the null hypothesis $\rho_1 = \rho_2$ for two independent random samples using the χ^2 distribution. This method is of value in that it can also be used to test the homogeneity of more than two sample values of r. For example, if we have several independent random samples, each with a given value of r, we may wish to test the null hypothesis $\rho_1 = \rho_2 = \cdots = \rho_k$. We can make this test in the case of $k = 2$ samples by finding

$$\chi^2 = (n_1 - 3)z_1^2 + (n_2 - 3)z_2^2 - \frac{[(n_1 - 3)z_1 + (n_2 - 3)z_2]^2}{(n_1 - 3) + (n_2 - 3)} \tag{8.8}$$

with $k - 1 = 1$ d.f. For our two samples, we have

$$\chi^2 = (15 - 3)(1.472)^2 + (10 - 3)(1.293)^2$$
$$- \frac{[(15 - 3)(1.472) + (10 - 3)(1.293)]^2}{(15 - 3) + (10 - 3)}$$

$$= 37.704 - \frac{(26.715)^2}{19}$$

$$= .141$$

and we find, from the table of χ^2, Table II in the Appendix, that with 1 d.f., the probability of obtaining $\chi^2 \geq .141$, when the null hypothesis is true, is approximately .70.

Because Z as defined by (8.6) is a standard normal variable, and because the square of a standard normal variable is a value of χ^2 with 1 d.f., the χ^2 test is the equivalent of a two-sided test using the standard normal distribution. In our example, we note that $Z^2 = (.376)^2 = .141$, which is equal to $\chi^2 = .141$.

8.10 Test for Homogeneity of Several Values of r

As we have stated previously, the χ^2 test is applicable to the case of $k > 2$ independent random values of r. Table 8.2 gives the sample sizes and values of r for each of $k = 4$ independent random samples. The z_r values for each value of r are given in column (4). Column (5) gives the values of $(n_i - 3)z_i$ and column (6) gives the values of $(n_i - 3)z_i^2$. In general, for k independent values of r,

$$\chi^2 = \sum_1^k (n_i - 3)z_i^2 - \frac{\left[\sum_1^k (n_i - 3)z_i\right]^2}{\sum_1^k (n_i - 3)} \tag{8.9}$$

with $k - 1$ d.f.

Substituting the appropriate values from Table 8.2 in (8.9), we have

$$\chi^2 = 73.126 - \frac{(109.140)^2}{168} = 2.22$$

TABLE 8.2 Calculation of the χ^2 test of the homogeneity of $k = 4$ independent values of r

	(1) n_i	(2) r_i	(3) $n_i - 3$	(4) z_i	(5) $(n_i - 3)z_i$	(6) $(n_i - 3)z_i^2$
	33	.53	30	.590	17.700	10.443
	58	.62	55	.725	39.875	28.909
	42	.65	39	.775	30.225	23.424
	47	.45	42	.485	21.340	10.350
Σ	180	2.25	168	2.575	109.140	73.126

a nonsignificant value with $k - 1 = 3$ d.f. and with $\alpha = .05$. We thus conclude that the various r's can be considered estimates of the same population value ρ. Therefore, we calculate

$$\bar{z}_r = \frac{\sum_{1}^{k}(n_i - 3)z_i}{\sum_{1}^{k}(n_i - 3)} \tag{8.10}$$

or the weighted average value of z_r. For the data of Table 8.2, we have

$$\bar{z}_r = \frac{109.140}{168} = .650$$

From Table VI in the Appendix we find that the corresponding value of r is approximately .570. We may regard this value as an estimate of the common population value ρ based on the combined data from the four samples.

8.11 Test of Significance of a Partial Correlation Coefficient: $r_{12.3}$

If we have three variables and we know the values of their intercorrelations, then we have defined the partial correlation coefficient between X_1 and X_2 with X_3 held constant as

$$r_{12.3} = \frac{r_{12} - r_{13}r_{23}}{\sqrt{1 - r_{13}^2}\sqrt{1 - r_{23}^2}}$$

To test the null hypothesis that $\rho_{12.3} = 0$, we may either calculate

$$t = \frac{r_{12.3}}{\sqrt{1 - r_{12.3}^2}}\sqrt{n - 3} \tag{8.11}$$

with $n - 3$ d.f. or use Table V in the Appendix with $n - 3$ d.f. rather than $n - 2$ d.f.

Exercises

8.1. For a random sample of $n = 10$, we have $r = .88$. Test the null hypothesis that $\rho = 0$. Make a two-sided test with $\alpha = .05$.

8.2. With a sample of $n = 20$, what value of r would be required in order to reject the null hypothesis that $\rho = 0$, if a two-sided test is made with $\alpha = .05$?

8.3. For a sample of $n = 25$, we have $r = .82$. Find a 95 percent confidence interval for ρ.

8.4. The correlation coefficient between scores on an algebra test and grades on a final examination for one section of an elementary statistics class consisting of 48 students was .56. For another section of 44 students the correlation coefficient was .45. Do these two values of r differ significantly? Make the test using both Z and χ^2. You should find that $Z^2 = \chi^2$.

8.5. For three independent random samples the correlation coefficients between scores on two personality scales were .83, .70, and .90. The number of observations on which the correlation coefficients were based were 28, 39, and 52, respectively. (a) Determine whether the correlation coefficients differ significantly, with $\alpha = .05$. (b) If they do not differ significantly, what is the estimated value of ρ based on the combined samples?

8.6. Briefly define each of the following concepts or terms:

confidence interval — significance level of a test

Type I error — null hypothesis

bivariate normal distribution

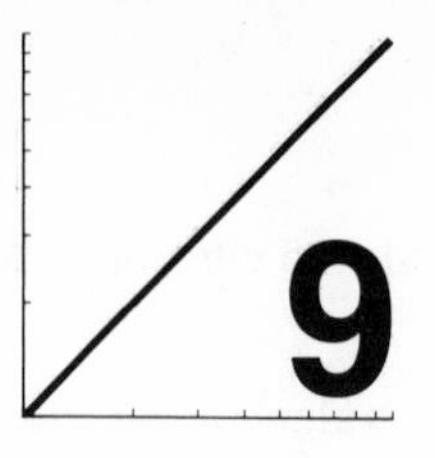

9

Tests of Significance for Special Cases of the Correlation Coefficient

9.1 Introduction

When X and Y have, in the population, a bivariate normal distribution, then we have seen that it is possible to make various tests of significance regarding the population value ρ. For example, we could test a null hypothesis $\rho = .50$ for a single sample or a null hypothesis $\rho_1 = \rho_2$ for two independent random samples, or that $\rho_1 = \rho_2 = \cdots = \rho_k$ for k independent random samples.

In the special cases of the correlation coefficient, described in Chapter 7, the joint distribution of X and Y cannot be a bivariate normal distribution. This is obviously impossible in the case of the correlation coefficient between two dichotomous variables, and in the case of a dichotomous variable and another variable that can take any number of different values. It is also impossible in the case of the correlation coefficient between two sets of ranks. This means that, in general, the various tests of significance we have described with respect to correlation coefficients for bivariate normally distributed variables do not apply in the special cases of the correlation coefficient described in Chapter 7.

We can, however, show that if the Y values are independent of the X values, then the phi coefficient, the point biserial coefficient, and the rank order correlation coefficient must, of necessity, be equal to zero. Thus any positive or negative value of r for any of these three special cases may be assumed to offer evidence that the Y values are not independent of the X values. If the obtained value of r in these special cases is sufficiently large that we reject the null hypothesis of independence, then we may conclude that the Y values are dependent on the X values or, in other words, that there is a significant association between the X and Y values.

9.2 Test of Significance for the Phi Coefficient

We consider first the correlation coefficient between two dichotomous variables. It will simplify the discussion to consider two items, X and Y, the response to each of which may be either correct or incorrect. We assign a value of 1 to a correct response and a value of 0 to an incorrect response. Figure 9.1(a) shows, for a sample of $n = 200$ individuals, $p_1 = .50$, the proportion giving the correct response to Item 1 (X), and $p_2 = .60$, the proportion giving the correct response to Item 2 (Y). We note that $q_1 = 1 - p_1$ and that $q_2 = 1 - p_2$. The cells of the figure give the possible response patterns (1,0), (1,1), (0,0), and (0,1). If the sample of subjects is random, then p_1, which is the mean value of X, can be shown to be an unbiased estimate of the population mean P_1, and P_1 is simply the probability that $X = 1$ if an observation is drawn at random from the population of X values. Similarly, p_2 is an unbiased

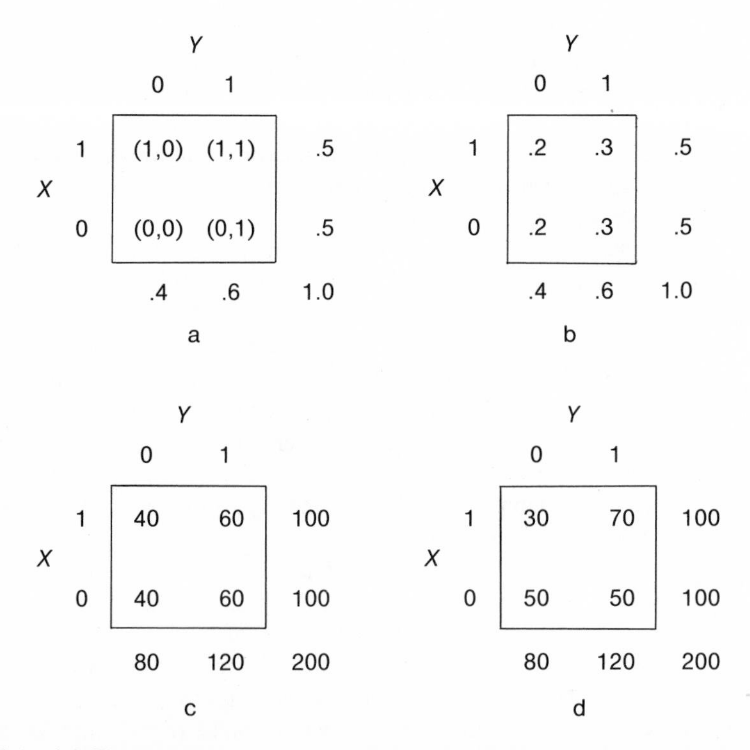

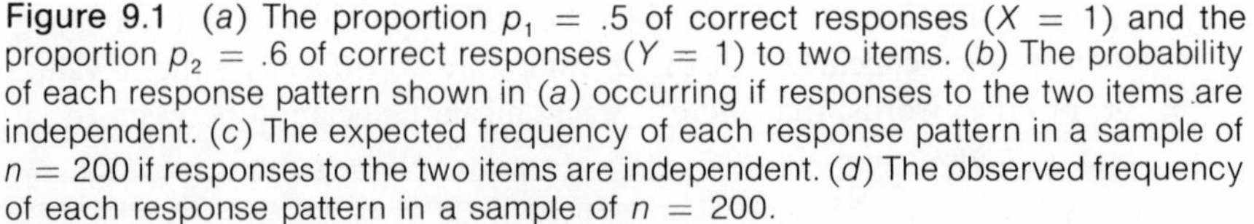
Figure 9.1 (a) The proportion $p_1 = .5$ of correct responses ($X = 1$) and the proportion $p_2 = .6$ of correct responses ($Y = 1$) to two items. (b) The probability of each response pattern shown in (a) occurring if responses to the two items are independent. (c) The expected frequency of each response pattern in a sample of $n = 200$ if responses to the two items are independent. (d) The observed frequency of each response pattern in a sample of $n = 200$.

estimate of P_2 and P_2 is simply the probability that $Y = 1$ if an observation is drawn at random from the population of Y values.

We recall from elementary algebra that if we have two independent events A and B and if $P(A)$ is the probability of A occurring and $P(B)$ is the probability of B occurring, then the probability that both A and B will occur will be given by the product of the two probabilities, that is $P(A)P(B)$. Similarly, if response to Item 1 is independent of response to Item 2, then the probability that $X = 1$ *and* $Y = 1$ or that the response pattern (1,1) will occur will be given by $P_1 \times P_2$. We have, as sample estimates of P_1 and P_2, $p_1 = .50$ and $p_2 = .60$. Thus, if response to Item 1 is independent of response to Item 2, we have $p_1 p_2 = (.5)(.6) = .30$ as an estimate of the probability of the response pattern (1,1), $p_1 q_2 = (.5)(.4) = .20$ as an estimate of the probability of the response pattern (1,0), $q_1 p_2 = (.5)(.6) = .30$ as an estimate of the probability of the response pattern (0,1), and $q_1 q_2 = (.5)(.4) = .20$ as an estimate of the probability of the response pattern (0,0). These probabilities have been entered in the cells of Figure 9.1(*b*).

If we multiply the probabilities just obtained by the sample size, $n = 200$, we obtain the *expected* number of individuals who will have each of the response patterns; these frequencies or expected numbers are shown in Figure 9.1(*c*). If we now calculate the correlation coefficient for the entries in Figure 9.1(*c*), we obtain

$$r = \frac{(60)(40) - (40)(60)}{\sqrt{(100)(100)(80)(120)}} = 0$$

Thus, if r is equal to zero, the observed data are in accord with the notion that responses to the two items are independent. On the other hand, any positive or negative value of r will indicate that the responses to the two items are not independent, but are instead associated.

Now, even when the responses to the two items in the population are independent, the observed cell entries corresponding to the possible patterns of response will show some deviation from the expected frequencies, simply as a result of random sampling. We are interested in determining whether the deviations are sufficiently large to reject the hypothesis that, in the population, responses to the two items are independent.

Assume, for example, that the observed frequencies are those shown in Figure 9.1(d) for a sample of $n = 200$. According to the hypothesis of independence, these observed frequencies, f_i, should deviate from the theoretical frequencies, F_i, only as a result of random sampling. As a test of the null hypothesis of independence, we calculate

$$\chi^2 = \sum \frac{(f_i - F_i)^2}{F_i} \tag{9.1}$$

and if the null hypothesis is true, χ^2 as obtained from (9.1) will be distributed as χ^2 with 1 d.f. If the obtained value of χ^2 exceeds the tabled value at, say, the $\alpha = .05$ significance level, the null hypothesis will be rejected. For the data in Figure 9.1(d), we have

$$\chi^2 = \frac{(30 - 40)^2}{40} + \frac{(70 - 60)^2}{60} + \frac{(50 - 40)^2}{40} + \frac{(50 - 60)^2}{60}$$

$$= 2.50 + 1.67 + 2.50 + 1.67$$

$$= 8.34$$

with 1 d.f. From the table of χ^2 we find that with 1 d.f., $P(\chi^2 \geq 3.841) = .05$, and we may conclude that the disagreement between the theoretical frequencies and the observed frequencies is such that the responses to the two items are not independent.

The χ^2 given by (9.1) is very simply related to the square of the correlation coefficient for two dichotomous variables of the kind described. In this particular case, it can be shown that

$$\chi^2 = nr^2 \tag{9.2}$$

In our example, we have

$$r^2 = \frac{[(70)(50) - (30)(50)]^2}{(80)(120)(100)(100)} = \frac{1}{24} = .0417$$

and, therefore,

$$\chi^2 = (200)(.0417) = 8.34$$

which is the same value we obtained using (9.1).

9.3 The *t* Test of Significance for the Point Biserial Coefficient

Table 9.1 gives the values of Y and Y^2, for a variable that can take a number of different values, in a sample in which $n_0 = 10$ observations have a value of $X = 0$ and $n_1 = 10$ observations have a value of $X = 1$. To determine the value of the point biserial coefficient, we first find

$$SS_T = n_0(\overline{Y}_0 - \overline{Y})^2 + n_1(\overline{Y}_1 - \overline{Y})^2$$

$$= \frac{(\Sigma Y_0)^2}{n_0} + \frac{(\Sigma Y_1)^2}{n_1} - \frac{(\Sigma Y_0 + \Sigma Y_1)^2}{n_0 + n_1} \tag{9.3}$$

or, for the data of Table 9.1,

$$SS_T = \frac{(30)^2}{10} + \frac{(80)^2}{10} - \frac{(110)^2}{20} = 125$$

We also have

$$SS_{tot} = 770 - \frac{(110)^2}{20} = 165$$

TABLE 9.1 Values of Y and Y^2 for a variable that can take a number of different values in a sample in which $n_0 = 10$ observations have a value of $X = 0$ and $n_1 = 10$ observations have a value of $X = 1$

	Values of Y		Values of Y^2	
	$X = 0$	$X = 1$	$X = 0$	$X = 1$
	4	8	16	64
	3	10	9	100
	3	10	9	100
	1	9	1	81
	2	6	4	36
	5	7	25	49
	5	7	25	49
	4	9	16	81
	1	6	1	36
	2	8	4	64
Σ	30	80	110	660

Then, for the value of the point biserial coefficient, we obtain

$$r = \sqrt{\frac{SS_T}{SS_{tot}}} = \sqrt{\frac{125}{165}} = \sqrt{.7576} = .8704$$

To test the null hypothesis that in the population ρ is equal to zero, we can calculate

$$t = \frac{r}{\sqrt{1 - r^2}} \sqrt{n - 2} \tag{9.4}$$

In our example we have

$$t = \frac{.8704}{\sqrt{1 - .7576}} \sqrt{20 - 2} = 7.50$$

with 18 d.f. With 18 d.f., we find from the table of t that $P(t \geq 7.50)$ or $P(t \leq -7.50)$ is less than .01, and the null hypothesis is thus rejected.

We could also use Table V in the Appendix to determine whether $r = .8704$ is significant. For a two-sided test, with $\alpha = .01$, and with 18 d.f., we find that either $r \geq .561$ or $r \leq -.561$ would result in the rejection of the null hypothesis. Because our obtained value of the point biserial coefficient, .8704, exceeds the tabled value, the null hypothesis would be rejected.

In one of the formulas we developed for the point biserial coefficient, we showed that if $\overline{Y}_0$ is equal to $\overline{Y}_1$, then r must be equal to zero. If the sample is random, then $\overline{Y}_0$ is an unbiased estimate of the population mean μ_0 and $\overline{Y}_1$ is an unbiased estimate of the population mean μ_1. In the case of a point biserial coefficient, the t test of the null hypothesis that $\rho = 0$, as given by (9.4), is equivalent to a test of the null hypothesis that in the population $\mu_0 = \mu_1$. The only assumptions involved in the t test for the point biserial coefficient are that the Y values for each value of X are normally distributed about the corresponding population means and that the population variance of the Y values is the same for each of the X values.

Under the assumption that $\sigma^2_{Y.0} = \sigma^2_{Y.1} = \sigma^2_Y$, our best estimate of the common population variance will be given by

$$s^2_Y = \frac{\Sigma y_0^2 + \Sigma y_1^2}{n_0 + n_1 - 2} \tag{9.5}$$

For the data of Table 9.1, we have

$$\Sigma y_0^2 = 110 - \frac{(30)^2}{10} = 20$$

and

$$\Sigma y_1^2 = 660 - \frac{(80)^2}{10} = 20$$

Then

$$s^2_Y = \frac{20 + 20}{10 + 10 - 2} = 2.2222$$

If the null hypothesis $\mu_0 = \mu_1$ is true, then

$$t = \frac{\overline{Y}_1 - \overline{Y}_0}{\sqrt{s^2_Y\left(\frac{1}{n_0} + \frac{1}{n_1}\right)}} \tag{9.6}$$

will have a t distribution with $n_0 + n_1 - 2$ d.f. Substituting $\overline{Y}_1 = 8$, $\overline{Y}_0 = 3$, $s^2_Y = 2.2222$, and $n_0 = n_1 = 10$ in (9.6), we obtain

$$t = \frac{8 - 3}{\sqrt{2.2222\left(\frac{1}{10} + \frac{1}{10}\right)}} = 7.50$$

which is equal to the value of t we obtained using (9.4). We conclude that $\mu_0 \neq \mu_1$ or, in other words, that the values of the Y means are not independent of the X classification.

9.4 The F Test of Significance for the Point Biserial Coefficient

A test of the null hypothesis $\mu_0 = \mu_1$ may also be made in terms of the F distribution. If the null hypothesis is true, then it can be shown that

$$F = \frac{SS_T}{s^2_Y} \tag{9.7}$$

where s^2_Y is defined by (9.5), will be distributed as F with 1 d.f. for the numerator and $n_0 + n_1 - 2$ d.f. for the denominator.

In our example, we have $SS_T = 125$ and $s_{\bar{Y}}^2 = 2.2222$. Then

$$F = \frac{125}{2.2222} = 56.25$$

with 1 and 18 d.f. From the table of F, Table VII in the Appendix, we find that for 1 and 18 d.f., $P(F \geq 56.25)$ is less than .01 and the null hypothesis would be rejected.

We note that the square of t, as defined by (9.6), will always be equal to F, as defined by (9.7). In our example, we have $t^2 = (7.50)^2 = 56.25$ and this is equal to the value of F.

Regardless of whether a t test or an F test is made, a significant value indicates that the values of the Y means, $\bar{Y}_0$ and $\bar{Y}_1$, are not independent of the X classification.

9.5 Test of Significance for the Rank Order Correlation Coefficient

The value of the rank order correlation coefficient can be regarded as indicating whether one set of ranks is independent of the other set of ranks or whether there is some agreement between the two sets of rankings. The distribution of the rank order correlation coefficient could, in theory, be calculated for any two sets of n ranks. We could, for example, arrange the ranks for one variable in standard order from 1 to n. We could then take every possible permutation of the ranks for the other variable and correlate each of the resulting permutations with the standard order for the first set of ranks.[1] According to the null hypothesis that the two sets of ranks are independent, each of these permutations is equally likely.

Olds[2] tabled the values of ΣD^2 for n ranks from 2 through 7 in terms of exact frequencies in the manner just described and for n equal to 8, 9, and 10 by means of an approximation function. We have used his table to calculate the corresponding values of r. These values are given in Table IX in the Appendix, which enables us to test the null hypothesis that the ranks are independent for relatively small values of n. The first column of Table IX gives the value of n, the number of ranks from 4 to 10, and the second column gives selected values of r for these sets of ranks. The third column gives the probability of obtaining r equal to or greater than the value given in column 2, when the null hypothesis is true. The tabled values of r are thus for a one-sided test. For example, with $n = 7$, the probability of $r \geq .679$ is equal to .0548, and this is

[1] We, in essence, regard one set of ranks as fixed and the other as random.

[2] Olds, E. G. Distributions of sums of rank differences for small numbers of individuals. *Annals of Mathematical Statistics*, 1938, *9*, 133–148.

also the probability of obtaining $r \leq -.679$, when the null hypothesis of independence is true. The probability of $r \geq .679$ or $r \leq -.679$ for a two-sided test will be equal to $(2)(.0548) = .1096$.

When n is greater than 10, the sampling distribution of the rank order coefficient, under the null hypothesis that $\rho = 0$, may be approximated by the t distribution. Thus

$$t = \frac{r}{\sqrt{1 - r^2}} \sqrt{n - 2}$$

with $n - 2$ d.f. provides a test of the null hypothesis that $\rho = 0$. It is, however, not necessary to calculate t to determine whether r is significant. Instead, we may use Table V in the Appendix, with $n - 2$ d.f., and evaluate the rank order correlation coefficient in terms of the tabled values in the same manner in which this table was used to test the significance of the point biserial coefficient.

Exercises

9.1. In a test consisting of two items, the proportion answering True to Item 1 is .60 and the proportion answering True to Item 2 is .70. (a) If responses to the two items are independent, what is the estimated probability of each of the response patterns (1,0), (1,1), (0,0), and (0,1) occurring? (b) If the sample consists of $n = 50$ individuals, what are the expected frequencies of each of the response patterns? (c) If the observed frequency of the response pattern (1,1) is 28, will responses to the two items be judged to be independent?

9.2. The scores of eight males and eight females on a test (Y) are as follows:

Males	Females
7	7
5	7
6	10
3	6
4	5
4	4
1	8
2	9

(a) Use a t test to determine whether the Y means are independent of the sex classification. (b) Use an F test to determine whether the Y means are independent of the sex classification. Note that $F = t^2$.

9.3. If the rank order correlation coefficient for a sample of $n = 8$ is equal to .80, would the null hypothesis $\rho = 0$ be rejected for a two-sided test with $\alpha = .05$?

9.4. For a sample of $n = 200$ the phi coefficient between responses to Item 1 and Item 2 is .23. Are responses to the two items independent?

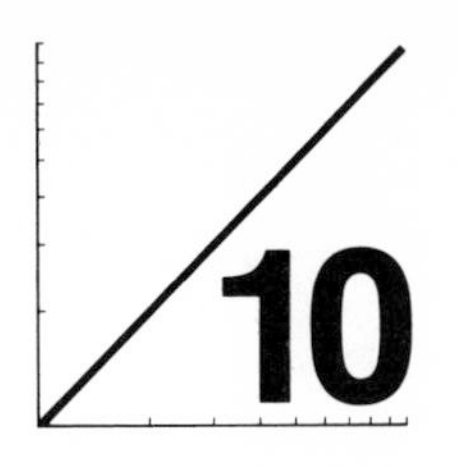

Tests of Significance for Regression Coefficients

10.1 Introduction

As we have pointed out previously, in an experiment the values of X, the independent variable, are usually preselected by the experimenter and may be regarded as fixed. The X values, in other words, are not subject to random variation. For each fixed value of X we have a random sample of one or more observations of a dependent variable Y. If we have calculated a regression coefficient b_Y for a set of paired (X,Y) values, where the X values are fixed, we may be interested in determining whether the obtained value of b_Y differs significantly from zero; that is, we may wish to test the null hypothesis that the population parameter β is equal to zero.[1]

In this chapter we show how the t distribution may be used to test various null hypotheses regarding β and also to test the null hypothesis that $\beta_1 - \beta_2 = 0$, if we have two independent samples of paired (X,Y) values. In addition, we shall see how to apply a test of significance to determine whether three or more sample regression coefficients based on independent random samples can be assumed to be estimates of a common population value. The only distribution assumptions of these tests of significance for regression coefficients are those involving the Y variable.

It is assumed that there is a normally distributed population of Y values with the same variance $\sigma^2_{Y.X}$ for each fixed value of X. The ordered values of $\mu_{Y.X}$, the population means of the Y values for each of the X values, are assumed to fall on a straight line with slope equal to β. Note that the assumptions of tests of significance of regression coefficients are considerably less restrictive than those of tests of significance of correlation coefficients. We do not, for example, have to assume that the joint distribution of X and Y is bivariate normal.

---

[1]In this chapter we shall be concerned only with the regression coefficient b_Y and the subscript will be omitted in the discussion that follows.

In regression problems in which the X values are fixed, the relationship between the Y and X values may take forms other than a linear relationship. Methods for studying other than linear relationships will be discussed in Chapters 11 and 12. In this chapter we are concerned solely with tests of significance of regression coefficients for linear relationships.

10.2 Test of the Null Hypothesis that $\beta = 0$

Table 10.1 gives the X and Y values for two independent random samples of $n_1 = 10$ and $n_2 = 10$ observations. For the first sample, we have

$$\Sigma(X - \overline{X})^2 = 385 - \frac{(55)^2}{10} = 82.5$$

$$\Sigma(Y - \overline{Y})^2 = 694 - \frac{(78)^2}{10} = 85.6$$

TABLE 10.1 Values of (X, Y) in two independent samples of $n = 10$ observations each

	Sample 1				
	X	Y	X^2	Y^2	XY
	1	2	1	4	2
	2	4	4	16	8
	3	8	9	64	24
	4	6	16	36	24
	5	8	25	64	40
	6	10	36	100	60
	7	8	49	64	56
	8	9	64	81	72
	9	12	81	144	108
	10	11	100	121	110
Σ	55	78	385	694	504
	Sample 2				
	X	Y	X^2	Y^2	XY
	1	6	1	36	6
	2	4	4	16	8
	3	8	9	64	24
	4	6	16	36	24
	5	8	25	64	40
	6	10	36	100	60
	7	12	49	144	84
	8	10	64	100	80
	9	11	81	121	99
	10	13	100	169	130
Σ	55	88	385	850	555

and

$$\Sigma(X - \overline{X})(Y - \overline{Y}) = 504 - \frac{(55)(78)}{10} = 75.0$$

Then, for the value of the regression coefficient, we obtain

$$b_1 = \frac{75.0}{82.5} = .91$$

and the residual sum of squares will be given by

$$\Sigma(Y - Y')^2 = 85.6 - \frac{(75.0)^2}{82.5} = 17.42$$

The residual variance is equal to

$$s_{Y.X}^2 = \frac{17.42}{10 - 2} = 2.18$$

The standard error of the regression coefficient will then be given by

$$s_b = \frac{s_{Y.X}}{\sqrt{\Sigma x^2}} \tag{10.1}$$

or

$$s_b = \sqrt{\frac{s_{Y.X}^2}{\Sigma x^2}} \tag{10.2}$$

and, in our example, we have

$$s_b = \sqrt{\frac{2.18}{82.5}} = \sqrt{.0264} = .1625$$

To test a null hypothesis regarding the value of β, we define

$$t = \frac{b - \beta}{s_b} \tag{10.3}$$

and if the null hypothesis is true, then the t defined by (10.3) will have a t distribution with $n - 2$ d.f. Specifically, if we wish to test the null hypothesis that $\beta = 0$, then, in our example, we have

$$t = \frac{.91}{.1625} = 5.60$$

with $10 - 2 = 8$ d.f. For a two-sided test of significance with 8 d.f. and with $\alpha = .05$, we find that $t \geq 2.306$ or $t \leq -2.306$ will be judged significant.

As we pointed out in our earlier discussion of the correlation coefficient, if b_Y is equal to zero, then r must also be equal to zero. The t test of the null hypothesis that $\beta = 0$ is the same as the t test of the null hypothesis that $\rho = 0$. For the sample under discussion, we have

$$r = \frac{75.0}{\sqrt{(82.5)(85.6)}} = .8925$$

and, as a test of the null hypothesis that $\rho = 0$, we have

$$t = \frac{.8925}{\sqrt{1 - (.8925)^2}} \sqrt{10 - 2} = 5.60$$

which is the same value of t we obtained when we tested the null hypothesis that $\beta = 0$. The proof that these two values of t must be equal is given in the answer to one of the exercises at the end of this chapter.

We see, therefore, that the t test of the null hypothesis $\rho = 0$ requires no assumptions other than those of the t test of the null hypothesis $\beta = 0$. This is also true of the t test of the null hypothesis $\rho = 0$ for a point biserial coefficient.

For the second sample of $n_2 = 10$ observations, we have

$$\Sigma(X - \overline{X})^2 = 385 - \frac{(55)^2}{10} = 82.5$$

$$\Sigma(Y - \overline{Y})^2 = 850 - \frac{(88)^2}{10} = 75.6$$

and

$$\Sigma(X - \overline{X})(Y - \overline{Y}) = 555 - \frac{(55)(88)}{10} = 71.0$$

Then, the regression coefficient for the second sample is

$$b_2 = \frac{71.0}{82.5} = .86$$

and for the sum of squared errors of prediction, we have

$$\Sigma(Y - Y')^2 = 75.6 - \frac{(71.0)^2}{82.5} = 14.5$$

For the residual variance for this sample, we have

$$s_{Y.X}^2 = \frac{14.5}{10 - 2} = 1.81$$

and for the standard error of b_2, we obtain

$$s_{b_2} = \sqrt{\frac{1.81}{82.5}} = \sqrt{.0219} = .1480$$

and

$$t = \frac{.86}{.1480} = 5.81$$

which is also a significant value with $\alpha = .05$ and with $10 - 2 = 8$ d.f.

10.3 Test of the Null Hypothesis that $\beta_1 - \beta_2 = 0$

We now wish to determine whether $b_1 = .91$ and $b_2 = .86$ differ significantly, that is, we wish to test the null hypothesis that $\beta_1 - \beta_2 = 0$. We assume homogeneity of the residual variances for the two groups on the Y variable and obtain as our estimate of the common residual variance[2]

$$s^2_{Y.X} = \frac{\left[\Sigma y_1^2 - \frac{(\Sigma xy_1)^2}{\Sigma x_1^2}\right] + \left[\Sigma y_2^2 - \frac{(\Sigma xy_2)^2}{\Sigma x_2^2}\right]}{n_1 + n_2 - 4} \tag{10.4}$$

Substituting the values we have already calculated in (10.4), we have

$$s^2_{Y.X} = \frac{\left[85.6 - \frac{(75.0)^2}{82.5}\right] + \left[75.6 - \frac{(71.0)^2}{82.5}\right]}{10 + 10 - 4}$$

$$= \frac{17.42 + 14.50}{16}$$

$$= 1.995$$

The standard error of the difference between the two regression coefficients will be given by

$$s_{b_1 - b_2} = \sqrt{s^2_{Y.X}\left(\frac{1}{\Sigma x_1^2} + \frac{1}{\Sigma x_2^2}\right)} \tag{10.5}$$

where $s^2_{Y.X}$ is defined by (10.4) and, in our example, is equal to 1.995. Substituting $\Sigma x_1^2 = \Sigma x_2^2 = 82.5$ in (10.5), we have

$$s_{b_1 - b_2} = \sqrt{1.995\left(\frac{1}{82.5} + \frac{1}{82.5}\right)} = .22$$

We then define

$$t = \frac{(b_1 - b_2) - (\beta_1 - \beta_2)}{s_{b_1 - b_2}} \tag{10.6}$$

The t defined by (10.6) will have a t distribution with $n_1 + n_2 - 4$ d.f., when the null hypothesis is true, and can be evaluated in terms of the table of t. In our example, we have $b_1 = .91$ and $b_2 = .86$ with $s_{b_1 - b_2} = .22$. As a test of the null hypothesis that $\beta_1 - \beta_2 = 0$, we have

$$t = \frac{.91 - .86}{.22} = .23$$

[2]We have $s^2_{Y.X} = 2.18$ for the first sample and $s^2_{Y.X} = 1.81$ for the second sample. Then, as a test of the validity of the assumption that the two variance estimates are homogeneous, we have $F = 2.18/1.81 = 1.20$ with 8 and 8 d.f., and this is a nonsignificant value of F with $\alpha = .01$.

From the table of t we find that for 16 d.f. and with a two-sided test, $t \geq 2.12$ or $t \leq -2.12$ will be significant with $\alpha = .05$, when the null hypothesis is true. Because our obtained value of t is equal to .23, we can regard the null hypothesis $\beta_1 - \beta_2 = 0$ as tenable.

The test of the null hypothesis that $\beta_1 - \beta_2 = 0$ is *not* equivalent to the test of the null hypothesis that $\rho_1 - \rho_2 = 0$. The test of the null hypothesis $\beta_1 - \beta_2 = 0$ is a test to determine whether the slopes of the two regression lines differ significantly. The slopes may differ significantly, whereas the two correlation coefficients may not. Similarly, the slopes of the two regression lines may not differ significantly, whereas the two correlation coefficients may differ significantly.

10.4 Test for Homogeneity of Several Independent Values of b

In some experiments, we may have three or more independent random samples. For each of the samples we may calculate the regression coefficient b. We are interested in determining whether the several sample values of b may be regarded as estimating the same common population value β. A test of the null hypothesis that $\beta_1 = \beta_2 = \cdots = \beta_k$ can be made in terms of the F distribution.

Table 10.2 gives the values of Σx^2, Σxy, Σy^2, b, $(\Sigma xy)^2/\Sigma x^2$, and

$$\Sigma(Y - Y')^2 = \Sigma y^2 - \frac{(\Sigma xy)^2}{\Sigma x^2}$$

for each of three independent samples of $n = 5$ observations each.[3]

The residual sums of squares $\Sigma(Y - Y')^2$ shown in column (7) of the table for each of the three samples have been minimized because for each sample we used the regression coefficient unique to the sample in obtaining the residual sum of squares. The sum of the three residual sums of squares is equal to 9.66, and this sum of squares is designated as SS_1 in the table. For SS_1 we have $k(n - 2)$ d.f., where k is the number of samples and n is the number of observations in each sample. Thus, in this problem, we have $3(5 - 2) = 9$ d.f. for SS_1.

Let us assume, for the moment, that the separate regression coefficients are all estimates of the same common population regression coefficient. Then an estimate of this common regression coefficient, which we designate as b_w, will be

$$b_w = \frac{\Sigma xy_1 + \Sigma xy_2 + \Sigma xy_3}{\Sigma x_1^2 + \Sigma x_2^2 + \Sigma x_3^2} \tag{10.7}$$

[3]The test of significance does not require that we have the same number of observations in each group.

TABLE 10.2 Values of Σx^2, Σxy, Σy^2, b, $(\Sigma xy)^2/\Sigma x^2$, and $\Sigma(Y - Y')^2$ for three independent samples of $n = 5$ observations each

(1) Sample	(2) Σx^2	(3) Σxy	(4) Σy^2	(5) b	(6) $(\Sigma xy)^2/\Sigma x^2$	(7) $\Sigma(Y - Y')^2$	(8) d.f.
(1) Sample 1	15.00	20.00	30.00	1.33	26.67	3.33	3
(2) Sample 2	15.00	23.00	38.00	1.53	35.27	2.73	3
(3) Sample 3	15.00	21.00	33.00	1.40	29.40	3.60	3
						$SS_1 = 9.66$	9
(4) Pooled	45.00	64.00	101.00	1.42	91.02	$SS_2 = 9.98$	11
						$SS_3 = SS_2 - SS_1 = 0.32$	2

or, in our example,

$$b_w = \frac{20.00 + 23.00 + 21.00}{15.0 + 15.0 + 15.0} = \frac{64.0}{45.0} = 1.42$$

Then we can also obtain a residual sum of squares representing the squared deviations $(Y - Y')^2$ when a line with a common slope equal to $b_w = 1.42$ is used for all three samples. We designate this residual sum of squares as SS_2 and, in our example,

$$SS_2 = 101.0 - \frac{(64.0)^2}{45.0} = 9.98$$

as shown in row (4) and column (7) of Table 10.2. SS_2 will have $k(n - 1) - 1$ d.f., or, in our example, $3(5 - 1) - 1 = 11$ d.f.[4]

Now SS_2 can never be smaller than SS_1, because SS_1 is based on the squared deviations within each sample from a regression line with slope b_i fitted separately for each sample, and the values of b_i were found in such a way as to minimize the sum of squared deviations within each sample. SS_2, on the other hand, is based on the squared deviations from a regression line that has the same slope, b_w, for each sample. Thus if the regression coefficient b_i for a given sample is not equal to b_w, the sum of squared deviations for this group from the line with slope b_w will be larger than that from the line with slope b_i. If $b_1 = b_2 = b_3 = b_w$, then SS_2 will be exactly equal to SS_1. If the b_i values do show considerable variation, then SS_2 will be considerably larger than SS_1.

To determine whether the regression coefficients differ significantly, we find

$$SS_3 = SS_2 - SS_1 \tag{10.8}$$

SS_3 will have $k - 1$ d.f., where k is the number of samples or separate regression coefficients. In our example, we have

$$SS_3 = 9.98 - 9.66 = .32$$

with $k - 1 = 2$ d.f.

For a test of significance of the differences among the regression coefficients, we have

$$F = \frac{SS_3/(k - 1)}{SS_1/k(n - 2)} \tag{10.9}$$

or, for the present problem,

[4]If the number of observations is not the same for the various samples, then the degrees of freedom for SS_1 will be given by $\Sigma_1^k n_k - 2k$, where n_k is the number of observations in a sample and k is the number of samples. For SS_2 the degrees of freedom will be given by $\Sigma_1^k n_k - k - 1$.

$$F = \frac{.32/2}{9.66/9} = \frac{.16}{1.07} = .15$$

with 2 and 9 d.f., and this is a nonsignificant value. We conclude that the data offer no significant evidence against the null hypothesis $\beta_1 = \beta_2 = \beta_3$. Our sample estimate of β is $b_w = 1.42$.

Exercises

10.1. We have a sample of $n = 5$ paired (X,Y) values as follows:

X	Y
5	10
4	6
3	4
2	4
1	1

(a) Calculate the value of the regression coefficient of Y on X. (b) Calculate $s^2_{Y.X}$ and the standard error s_b. (c) Test the null hypothesis that $\beta = 0$, using a two-sided t test with $\alpha = .05$.

10.2. We have another independent random sample of $n = 5$ paired (X,Y) values as follows:

X	Y
5	3
4	3
3	1
2	2
1	1

(a) Calculate the value of the regression coefficient of Y on X. (b) Calculate $s^2_{Y.X}$ and s_b. (c) Test the null hypothesis that $\beta = 0$ using a two-sided t test with $\alpha = .05$.

10.3. Use the data of exercises 10.1 and 10.2 and assume that $s^2_{Y.X}$ for each sample is an estimate of the same parameter $\sigma^2_{Y.X}$. (a) Pool the residual sums of squares and find $s^2_{Y.X}$ for the combined samples. (b) Find the standard error of the difference between the two regression coefficients. (c) Test the null hypothesis that $\beta_1 - \beta_2 = 0$, using a two-sided t test with $\alpha = .05$.

10.4. Use the two samples given in exercises 10.1 and 10.2. (a) Calculate SS_1, SS_2, and SS_3. (b) Use the F test to determine whether b_1 and b_2 differ significantly. (c) Show that the value of F obtained is equal to the square of the value of t obtained in exercise 10.3.

10.5. If one regression coefficient, b_1, based on one sample differs significantly from zero and another regression coefficient, b_2, based on an independent sample does not differ significantly from zero, does it necessarily follow that b_1 and b_2 will differ significantly? Explain your answer.

10.6. If both b_1 and b_2 for two independent samples differ significantly from zero, does it necessarily follow that the difference between b_1 and b_2 will not be significant? Explain your answer.

10.7. If b_Y is equal to zero, will r also be equal to zero? Explain why or why not.

10.8. Prove that the t test of the null hypothesis that $\rho = 0$, that is,

$$t = \frac{r}{\sqrt{1 - r^2}} \sqrt{n - 2}$$

is algebraically equivalent to the t test of the null hypothesis that $\beta = 0$, that is,

$$t = \frac{b}{\sqrt{s_{Y.X}^2 / \Sigma x^2}}$$

10.9. If we reject the null hypothesis $\beta_1 - \beta_2 = 0$ for two independent samples, would we also reject the null hypothesis $\rho_1 - \rho_2 = 0$, at the same significance level and for the same samples? Explain your answer.

10.10. What information is provided by the test of the null hypothesis $\beta_1 - \beta_2 = 0$ that is not provided by the test of the null hypothesis $\rho_1 - \rho_2 = 0$?

10.11. What information is provided by the test of the null hypothesis $\rho_1 - \rho_2 = 0$ that is not provided by the test of the null hypothesis $\beta_1 - \beta_2 = 0$?

10.12. If we have a typical correlation problem with a random sample of paired (X,Y) values, and if we reject the null hypothesis $\rho = 0$, would we also reject the null hypotheses $\beta_Y = 0$ and $\beta_X = 0$? Explain why or why not.

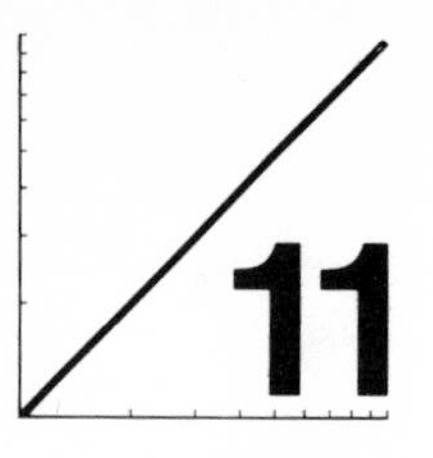

11 Coefficients for Orthogonal Polynomials

11.1 Introduction

Consider an experiment involving k equally spaced values of a quantitative variable X. For example, if X is a measure of time, the values of X might be 5 seconds, 10 seconds, 15 seconds, and 20 seconds, or they might be 2 minutes, 4 minutes, 6 minutes, and 8 minutes. In a learning experiment the values of X might consist of 1, 2, 3, 4, and 5 trials. If X is a dosage of a drug, then the values of X might be 1 grain, 2 grains, 3 grains, and so on. We assume that not only are the values of X equally spaced but that they are also fixed, that is, if the experiment were to be repeated, the values of X would be the same.

For each value of X we obtain n independent random observations of a dependent variable Y. The values of Y are assumed to be normally distributed for each value of X with the same variance $\sigma^2_{Y.X} = \sigma^2_{Y}$. The sample Y means obtained for each value of X are unbiased estimates of the corresponding population means. On the basis of the sample means we wish to infer something about the trend of the population means over the fixed values of X.

We let the ordered sums of the Y values be represented by $\Sigma Y_1, \Sigma Y_2, \ldots, \Sigma Y_k$ and the ordered means be represented by $\overline{Y}_1, \overline{Y}_2, \ldots, \overline{Y}_k$. The sum of all of the Y values will be represented by ΣY and the mean will be represented by $\overline{Y}$. Then the weighted sum of squared deviations of the means $\overline{Y}_i$ from the overall mean $\overline{Y}$ will be given by

$$SS_T = n\sum_{1}^{k}(\overline{Y}_i - \overline{Y})^2 \tag{11.1}$$

where SS_T is the treatment sum of squares with $k - 1$ d.f. An algebraic equivalent of (11.1) that may also simplify the calculations is

$$SS_T = \frac{(\Sigma Y_1)^2}{n} + \frac{(\Sigma Y_2)^2}{n} + \cdots + \frac{(\Sigma Y_k)^2}{n} - \frac{(\Sigma Y)^2}{kn} \qquad (11.2)$$

If the Y means all fall on a straight line, with slope $b_i \neq 0$, then all of the variation in the Y means can be accounted for by an equation of the first degree or a linear equation. That part of SS_T that can be accounted for by an equation of the first degree is referred to as the *linear component* of SS_T or as SS_L with 1 d.f. Then the residual sum of squares or the sum of squares for deviations from linearity will be given by

$$SS_{res} = SS_T - SS_L$$

with $k - 2$ d.f.

If SS_{res} represents significant deviations of the Y means from linearity, then we may determine how much of this residual variation can be accounted for by an equation of the second degree or by the *quadratic component* of SS_T. Any set of k means can always be described accurately by an equation of degree no higher than $k - 1$. With $k = 5$ means, for example, it is always possible to partition SS_T into a linear, quadratic, cubic, and quartic component corresponding to an equation of the first, second, third, and fourth degree. Not all of these components will necessarily be required to describe the trend of the means. Some components may be equal to zero or sufficiently small as to be nonsignificant.

In most behavioral science experiments, we would ordinarily be interested in first determining whether the linear component is significant, and then in determining whether the deviations from linearity are significant. If they are, then we would determine whether the quadratic component (SS_Q) is significant, and if so, whether there are significant deviations from the linear and quadratic components, that is, whether the new residual sum of squares

$$SS_{res} = SS_T - SS_L - SS_Q$$

is significant. Occasionally, we may also be interested in the cubic component and the quartic component.

11.2 Coefficients for Orthogonal Polynomials

Instead of working with polynomial equations of the second degree and higher, we make use of coefficients for orthogonal polynomials. These coefficients enable us to investigate systematically the trend of the means and to find out which components of the trend are significant using only an equation of the first degree. Table X in the Appendix[1] gives the coef-

[1]Table X gives only the coefficients for the linear, quadratic, cubic, and quartic components. Orthogonal coefficients for the higher-degree polynomials can be found in R. A. Fisher and F. Yates, *Statistical tables for biological, agricultural and medical research* (3rd ed.). Edinburgh: Oliver and Boyd, 1936.

TABLE 11.1 Orthogonal coefficients for the linear, quadratic, cubic, and quartic components of the trend for $k = 5$ means

Component		Coefficients				
Linear:	x_1	−2	−1	0	1	2
Quadratic:	x_2	2	−1	−2	−1	2
Cubic:	x_3	−1	2	0	−2	1
Quartic:	x_4	1	−4	6	−4	1

ficients for orthogonal polynomials for $k = 3$ to $k = 10$ values of X. Table 11.1 gives these coefficients for $k = 5$ values of X. Note that there are only $k - 1 = 4$ sets of orthogonal coefficients corresponding to the linear, quadratic, cubic, and quartic components. These coefficients are represented by x because, for each set of coefficients, we have

$$\Sigma x_1 = \Sigma x_2 = \Sigma x_3 = \Sigma x_4 = 0$$

The sets of coefficients are described as *orthogonal* because, for any two sets (for example, x_1 and x_2), we have

$$\Sigma x_1 x_2 = 0$$

Thus the correlation between x_1 and x_2 will be equal to zero, and this is also true for any two sets, x_i and x_j, $i \neq j$.

We have said that any set of k means can be accurately described by an equation of degree no higher than $k - 1$. With coefficients for orthogonal polynomials this means that the linear equation

$$\overline{Y}_i' = \overline{Y} + b_1x_1 + b_2x_2 + \cdots + b_{k-1}x_{k-1} \tag{11.3}$$

will result in a set of predicted values that are equal to the corresponding observed values, $\overline{Y}_i$, of the means. The successive terms $b_1x_1, b_2x_2, \ldots,$ $b_{k-1}x_{k-1}$ correspond to terms of the first, second, third, and higher degree in a polynomial equation. Not all of the terms may be needed to describe accurately the trend of the means. For example, the trend of the means may be described accurately by

$$\overline{Y}_i' = \overline{Y} + b_1x_1$$

or by an equation of the first degree. If both the linear and quadratic components of the trend are significant and the other components are not, then

$$\overline{Y}_i' = \overline{Y} + b_1x_1 + b_2x_2$$

may be assumed to describe the trend of the means. If only the quadratic component is significant, then

$$\overline{Y}_i' = \overline{Y} + b_2x_2$$

may be assumed to describe the trend of the means.

In (11.3) the values of $x_1, x_2, \ldots, x_{k-1}$ are known and can be obtained from Table X in the Appendix. The corresponding values of $b_1, b_2, \ldots, b_{k-1}$ are unknown, but they can be obtained from the data of the experiment.

In this chapter we will look at five hypothetical examples, each involving $k = 5$ means. In each example only *one* of the components is necessary to describe the trend of the means; that is, for all of the other components the value of b is equal to zero.

11.3 An Example in Which $\overline{Y}_i' = \overline{Y} + b_1 x_1$

Figure 11.1 shows a plot of $k = 5$ means against equally spaced values of an independent variable X. The ordered values of the five means are 4, 6, 8, 10, and 12, and each mean is based on $n = 5$ observations.[2] The means obviously fall on a straight line and we should find that the linear component of the trend is equal to SS_T. With equal n's, we have

$$\overline{Y} = \frac{4 + 6 + 8 + 10 + 12}{5} = 8$$

Then, using (11.1), we have

$$\begin{aligned} SS_T &= 5[(4-8)^2 + (6-8)^2 + (8-8)^2 + (10-8)^2 + (12-8)^2] \\ &= 200 \end{aligned}$$

Equivalently, using (11.2), we also have

$$SS_T = \frac{(20)^2}{5} + \frac{(30)^2}{5} + \frac{(40)^2}{5} + \frac{(50)^2}{5} + \frac{(60)^2}{5} - \frac{(200)^2}{25} = 200$$

To find the linear component of the trend, we multiply each ordered value of $\overline{Y}_i$ by the corresponding ordered value of x_1, the coefficient for the linear component. We have shown previously that

$$\Sigma(X - \overline{X})(Y - \overline{Y}) = \Sigma XY - \frac{(\Sigma X)(\Sigma Y)}{n}$$

[2]The coefficients for orthogonal polynomials given in Table X assume that we have equally spaced values of X and that we have the same number of observations of Y for each value of X. If the values of X are not equally spaced, or if the n's are not equal, the coefficients given in Table X are not applicable. Coefficients for orthogonal polynomials can be derived when the values of the independent variable are unequally spaced or when the n's are unequal, but it would obviously be impractical to table these coefficients for all possible cases. See J. Gaito, Unequal intervals and unequal n's in trend analyses, *Psychological Bulletin*, 1965, *63*, 125–127, for methods that can be used to derive the coefficients for unequal n's or for values of X that are not equally spaced.

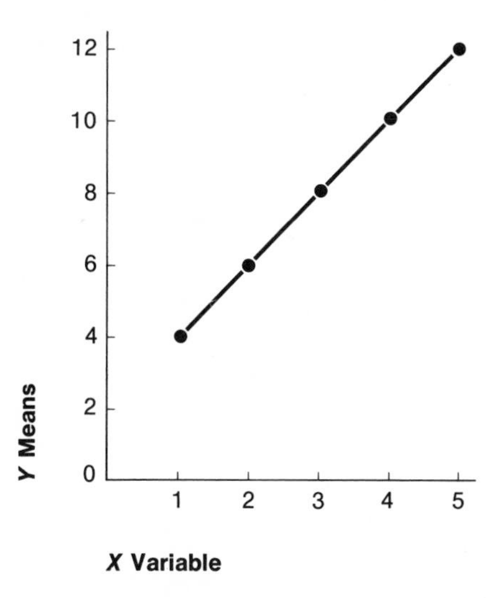

Figure 11.1 A plot of means in which $\overline{Y}_i' = \overline{Y} + b_1 x_1$ and where the values of x_1 are the orthogonal coefficients for the linear component of the trend.

The values of x_1 are in deviation form because $\Sigma x_1 = 0$. Thus

$$\begin{aligned}\Sigma x_1(\overline{Y}_i - \overline{Y}) &= \Sigma x_1 \overline{Y}_i - \overline{Y}\Sigma x_1 \\ &= \Sigma x_1 \overline{Y}_i\end{aligned}$$

and

$$\Sigma x_1 \overline{Y}_i = \frac{1}{n}\Sigma(x_1 \Sigma Y_i) \tag{11.4}$$

In our example, we have

$$\begin{aligned}\Sigma x_1 \overline{Y}_i &= \frac{1}{5}[(-2)(20) + (-1)(30) + (0)(40) + (1)(50) + (2)(60)] \\ &= \frac{1}{5}(100)\end{aligned}$$

The regression coefficient of $\overline{Y}_i$ on x_1 will then be

$$b_1 = \frac{\frac{1}{n}\Sigma(x_1 \Sigma Y_i)}{\Sigma x_1^2} = \frac{\Sigma(x_1 \Sigma Y_i)}{n\Sigma x_1^2} \tag{11.5}$$

or, in our example,

$$b_1 = \frac{100}{(5)(10)} = 2$$

Then the regression equation for predicting the Y means will be

$$\overline{Y}_i' = \overline{Y} + b_1 x_1$$

In our example, we have

$$\begin{aligned}
\overline{Y}_1' &= 8 + (2)(-2) = 4 \\
\overline{Y}_2' &= 8 + (2)(-1) = 6 \\
\overline{Y}_3' &= 8 + (2)(0) = 8 \\
\overline{Y}_4' &= 8 + (2)(1) = 10 \\
\overline{Y}_5' &= 8 + (2)(2) = 12
\end{aligned}$$

and we see that the values of $\overline{Y}_i$ are perfectly predicted by the regression equation $\overline{Y}_i' = \overline{Y} + b_1 x_1$.

The linear component of SS_T will be given by

$$SS_L = \frac{[\Sigma(x_1 \Sigma Y_i)]^2}{n \Sigma x_1^2} \tag{11.6}$$

or, in our example,

$$SS_L = \frac{(100)^2}{(5)(10)} = 200$$

We see that $SS_L = SS_T$. Consequently, for the sum of squared deviations from linearity, we have

$$SS_{res} = SS_T - SS_L = 0$$

Note also that the square of the correlation coefficient between x_1 and $\overline{Y}_i$ will be

$$\begin{aligned}
r^2_{x_1\overline{Y}_i} &= \frac{\left[\frac{1}{n}\Sigma(x_1 \Sigma Y_i)\right]^2}{(\Sigma x_1^2)[\Sigma(\overline{Y}_i - \overline{Y})^2]} \\
&= \frac{[\Sigma(x_1 \Sigma Y_i)]^2}{(n\Sigma x_1^2)[n\Sigma(\overline{Y}_i - \overline{Y})^2]}
\end{aligned}$$

but $n\Sigma(\overline{Y}_i - \overline{Y})^2$ is equal to SS_T. Therefore,

$$r^2_{x_1\overline{Y}_i} = \frac{[\Sigma(x_1 \Sigma Y_i)]^2}{(n\Sigma x_1^2)(SS_T)} \tag{11.7}$$

In our example, we have

$$r^2_{x_1\overline{Y}_i} = \frac{(100)^2}{(5)(10)(200)} = 1.00$$

Thus we see that all of the variation in the Y means can be accounted for in terms of the regression of the Y means on the values of x_1.

11.4 An Example in Which $\overline{Y} = \overline{Y} + b_2x_2$

Now consider the Y means shown in Figure 11.2. The ordered values of the Y means are 12, 6, 4, 6, and 12, and each mean is based on $n = 5$ observations. With equal n's we also have

$$\overline{Y} = \frac{12 + 6 + 4 + 6 + 12}{5} = 8$$

It is obvious, in this example, that there is no linear trend of the means and, consequently, the linear component will be equal to zero. The trend shown in the figure corresponds to a quadratic equation and we should find that the quadratic component of the trend is equal to SS_T. For the treatment sum of squares, using (11.2), we have

$$SS_T = \frac{(60)^2}{5} + \frac{(30)^2}{5} + \frac{(20)^2}{5} + \frac{(30)^2}{5} + \frac{(60)^2}{5} - \frac{(200)^2}{25} = 280$$

Using the coefficients x_2 for the quadratic component, we obtain

$$\frac{1}{n}\Sigma(x_2\Sigma Y_i) = \frac{1}{5}[(2)(60) + (-1)(30) + (-2)(20) + (-1)(30) + (2)(60)]$$
$$= \frac{1}{5}(140)$$

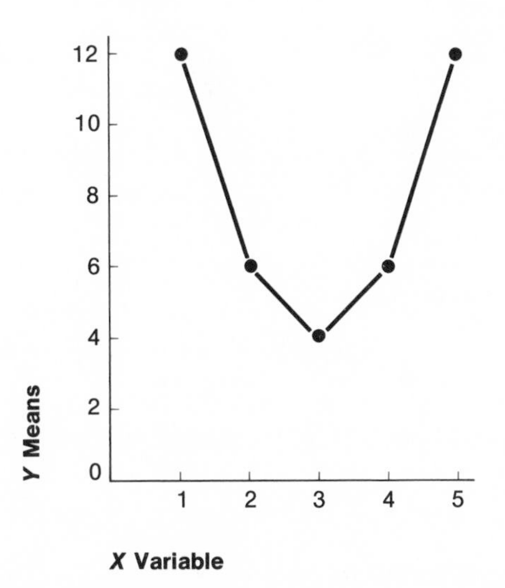

Figure 11.2 A plot of means in which $\overline{Y}_i' = \overline{Y} + b_2x_2$ and where the values of x_2 are the orthogonal coefficients for the quadratic component of the trend.

and

$$b_2 = \frac{\Sigma(x_2 \Sigma Y_i)}{n \Sigma x_2^2}$$

or, in our example,

$$b_2 = \frac{140}{(5)(14)} = 2$$

It can easily be determined that b_1 is equal to zero and, consequently,

$$\overline{Y}_i' = \overline{Y} + b_2 x_2$$

Then, we have

$$\overline{Y}_1' = 8 + (2)(2) = 12$$
$$\overline{Y}_2' = 8 + (2)(-1) = 6$$
$$\overline{Y}_3' = 8 + (2)(-2) = 4$$
$$\overline{Y}_4' = 8 + (2)(-1) = 6$$
$$\overline{Y}_5' = 8 + (2)(2) = 12$$

and we see that the Y means are perfectly predicted by the regression equation $\overline{Y}_i' = \overline{Y} + b_2 x_2$. For the square of the correlation coefficient between x_2 and $\overline{Y}_i$, we have

$$r^2_{x_2\overline{Y}_i} = \frac{[\Sigma(x_2 \Sigma Y_i)]^2}{(n \Sigma x_2^2)(SS_T)}$$

or, in our example,

$$r^2_{x_2\overline{Y}_i} = \frac{(140)^2}{(5)(14)(280)} = 1.00$$

The quadratic component of SS_T will be given by

$$SS_Q = \frac{[\Sigma(x_2 \Sigma Y_i)]^2}{n \Sigma x_2^2}$$

or, in our example,

$$SS_Q = \frac{(140)^2}{(5)(14)} = 280$$

which is equal to SS_T. All of the other components of SS_T can easily be shown to be equal to zero.

11.5 An Example in Which $\overline{Y}_i' = \overline{Y} + b_3 x_3$

For the $k = 5$ means, each based on $n = 5$ observations, shown in Figure 11.3, we have the ordered values 6, 12, 8, 4, and 10. The trend shown is

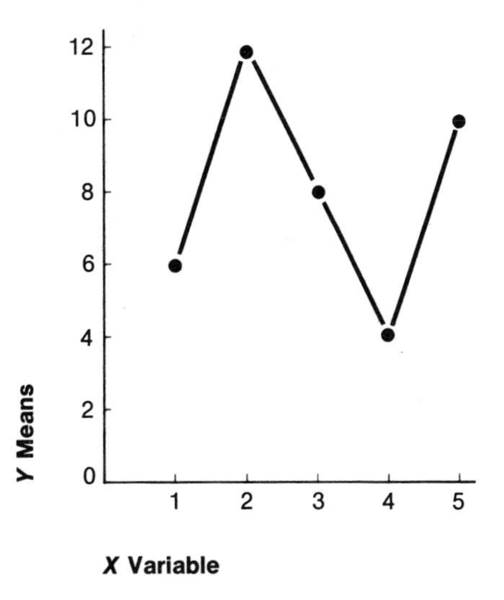

Figure 11.3 A plot of means in which $\overline{Y}_i' = \overline{Y} + b_3x_3$ and where the values of x_3 are the orthogonal coefficients for the cubic component of the trend.

that given by an equation of the third degree. Consequently, we should find that the regression equation

$$\overline{Y}_i' = \overline{Y} + b_3x_3$$

will predict perfectly the values of the Y means.

Multiplying the treatment means by the coefficients x_3 for the cubic component, we have

$$\frac{1}{n}\Sigma(x_3\Sigma Y_i) = \frac{1}{5}[(-1)(30) + (2)(60) + (0)(40) + (-2)(20) + (1)(50)]$$

$$= \frac{1}{5}(100)$$

Then

$$b_3 = \frac{\Sigma(x_3\Sigma Y_i)}{n\Sigma x_3^2}$$

and, in our example,

$$b_3 = \frac{100}{(5)(10)} = 2$$

Because all values of b other than b_3 are equal to zero, we have

$$\overline{Y}_i' = \overline{Y} + b_3x_3$$

and, in our example,

$$\overline{Y}_1' = 8 + (2)(-1) = 6$$
$$\overline{Y}_2' = 8 + (2)(2) = 12$$
$$\overline{Y}_3' = 8 + (2)(0) = 8$$
$$\overline{Y}_4' = 8 + (2)(-2) = 4$$
$$\overline{Y}_5' = 8 + (2)(1) = 10$$

and we see that the Y means are perfectly predicted by the regression equation $\overline{Y}_i' = \overline{Y} + b_3 x_3$.

For the sum of squares for the cubic component, we have

$$SS_C = \frac{[\Sigma(x_3 \Sigma Y_i]^2}{n \Sigma x_3^2}$$

or, in our example,

$$SS_C = \frac{(100)^2}{(5)(10)} = 200$$

and SS_C should be equal to SS_T. For SS_T, we have

$$SS_T = \frac{(30)^2}{5} + \frac{(60)^2}{5} + \frac{(40)^2}{5} + \frac{(20)^2}{5} + \frac{(50)^2}{5} - \frac{(200)^2}{25} = 200$$

We also have

$$r^2_{x_3 \overline{Y}_i} = \frac{[\Sigma(x_3 \overline{Y}_i)]^2}{(n \Sigma x_3^2)(SS_T)} = \frac{(100)^2}{(5)(10)(200)} = 1.00$$

11.6 An Example in Which $\overline{Y}_i' = \overline{Y} + b_4 x_4$

Figure 11.4 illustrates a case where the trend of the means can be accounted for in terms of the quartic component. Each of the means is based on $n = 5$ observations and the ordered values of the means are 8, 3, 13, 3, and 8. The overall mean $\overline{Y}$ is equal to 7. For the treatment sum of squares, we have

$$SS_T = \frac{(40)^2}{5} + \frac{(15)^2}{5} + \frac{(65)^2}{5} + \frac{(15)^2}{5} + \frac{(40)^2}{5} - \frac{(175)^2}{25} = 350$$

Using the coefficients x_4 for the quartic component, we have

$$\frac{1}{n}\Sigma(x_4 \Sigma Y_i) = \frac{1}{5}[(1)(40) + (-4)(15) + (6)(65) + (-4)(15) + (1)(40)]$$

$$= \frac{1}{5}(350)$$

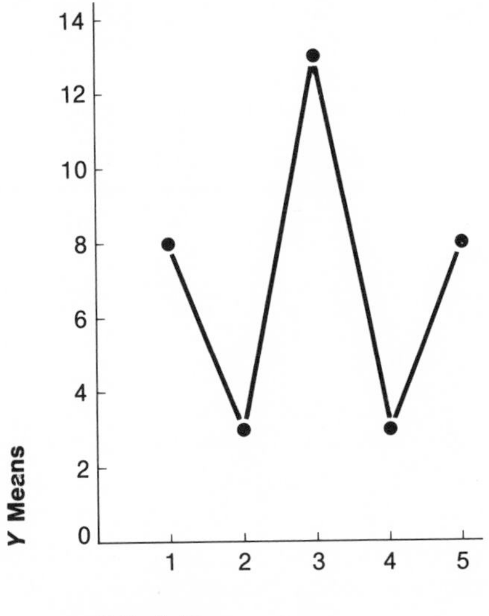

Figure 11.4 A plot of means in which $\overline{Y}_i' = \overline{Y} + b_4x_4$ and where the values of x_4 are the orthogonal coefficients for the quartic component of the trend.

Then

$$b_4 = \frac{\Sigma(x_4 \Sigma Y_i)}{n\Sigma x_4^2}$$

or, in our example,

$$b_4 = \frac{350}{(5)(70)} = 1$$

Because all values of b except b_4 are equal to zero, we have

$$\overline{Y}_i' = \overline{Y} + b_4x_4$$

or

$$\overline{Y}_1' = 7 + (1)(1) = 8$$
$$\overline{Y}_2' = 7 + (1)(-4) = 3$$
$$\overline{Y}_3' = 7 + (1)(6) = 13$$
$$\overline{Y}_4' = 7 + (1)(-4) = 3$$
$$\overline{Y}_5' = 7 + (1)(1) = 8$$

and we see that the observed values of $\overline{Y}_i$ are perfectly predicted by the regression equation $\overline{Y}_i' = \overline{Y} + b_4x_4$. Consequently,

$$r^2_{x_4\overline{Y}_i} = \frac{[\Sigma(x_4\overline{Y}_i)]^2}{(n\Sigma x_4^2)(SS_T)} = \frac{(350)^2}{(5)(70)(350)} = 1.00$$

For the sum of squares for the quartic component, we have

$$\frac{[\Sigma(x_4\Sigma Y_i)]^2}{n\Sigma x_4^2} = \frac{(350)^2}{(5)(70)} = 350$$

which is equal to SS_T.

11.7 Summary

The four examples described in this chapter are hypothetical and were designed to illustrate cases in which all of the variation in the Y means can be accounted for by either the linear, quadratic, cubic, or quartic component of the trend. In these hypothetical examples all of the values of b, except one, were equal to zero in the equation

$$\overline{Y}_i' = \overline{Y} + b_1x_1 + b_2x_2 + b_3x_3 + b_4x_4$$

For the linear example, we had $b_2 = b_3 = b_4 = 0$ and

$$\overline{Y}_i' = \overline{Y} + b_1x_1$$

Because the values of $\overline{Y}_i$ were predicted perfectly by the corresponding values of $\overline{Y}_i'$, the residual sum of squares

$$SS_{res} = n\Sigma(\overline{Y}_i - \overline{Y}_i')^2 = SS_T - SS_L$$

was equal to zero.

Similarly, for the quadratic example, we had $b_1 = b_3 = b_4 = 0$, and

$$\overline{Y}_i' = \overline{Y} + b_2x_2$$

predicted perfectly the corresponding values of $\overline{Y}_i$. In this example,

$$SS_{res} = n\Sigma(\overline{Y}_i - \overline{Y}_i')^2 = SS_T - SS_Q$$

was also equal to zero. Similar considerations apply to the cubic and quartic examples.

In the next chapter, we consider a more realistic example of the kind that we might encounter in an actual experiment.

Exercises

11.1. We have $k = 5$ equally spaced values of a variable X with $n = 5$ observations of a dependent variable Y for each value of X. The ordered Y means are 3, 4, 5, 6, and 7. (a) Calculate SS_T and the linear component of SS_T. (b)

Calculate the values of $\overline{Y}_i' = \overline{Y} + b_1x_1$ and show that the correlation between $\overline{Y}_i'$ and $\overline{Y}_i$ is equal to 1.00.

11.2. Assume that the ordered values of the $k = 5$ means in exercise 11.1 are 7, 4, 3, 4, and 7. (a) Calculate SS_T and the quadratic component of SS_T. (b) Calculate the values of $\overline{Y}_i' = \overline{Y} + b_2x_2$ and show that the correlation between $\overline{Y}_i'$ and $\overline{Y}_i$ is equal to 1.00.

11.3. Assume that the ordered values of the $k = 5$ means in exercise 11.1 are 4, 7, 5, 3, and 6. (a) Calculate SS_T and the cubic component of SS_T. (b) Calculate the values of $\overline{Y}_i' = \overline{Y} + b_3x_3$ and show that the correlation between $\overline{Y}_i'$ and $\overline{Y}_i$ is equal to 1.00.

11.4. What properties do the coefficients for orthogonal polynomials possess? For example, what is true about Σx_i? What is true about $\Sigma x_i x_j$, $i \neq j$?

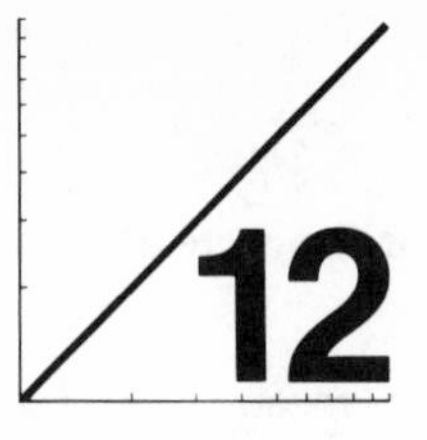

12 Tests of Significance Using Coefficients for Orthogonal Polynomials

12.1 Introduction

In this chapter we examine in some detail an experiment involving k equally spaced values of a quantitative variable X. There are n observations for each value of X. We first partition the total sum of squares (SS_{tot}) into the treatment sum of squares (SS_T) and the within treatment sum of squares (SS_W). Although we make a test of significance of the treatment mean square, the results of this test need not be significant to enable us to proceed with the subsequent tests of significance of the various components of the treatment sum of squares.

12.2 The Analysis of Variance for $k = 5$ Values of X

Table 12.1 gives the $n = 5$ values of Y for each of $k = 5$ equally spaced values of an independent variable X. In analyzing the data for this example, we first find the sum of squared deviations of the Y values from $\overline{Y}$, the overall mean of the Y values. This sum of squares is called the total sum of squares and will be given by

$$SS_{tot} = \Sigma(Y - \overline{Y})^2 = \Sigma Y^2 - \frac{(\Sigma Y)^2}{kn} \tag{12.1}$$

For the data of Table 12.1 we have

$$SS_{tot} = (10)^2 + (8)^2 + \cdots + (14)^2 - \frac{(300)^2}{(5)(5)} = 100$$

with $kn - 1$ d.f.

We then calculate the treatment sum of squares. In our example, we have

TABLE 12.1 Values of Y for $k = 5$ equally spaced values of a quantitative variable X

	X Variable				
	1	2	3	4	5
	10	12	9	14	15
	8	14	13	13	13
	12	11	12	11	12
	11	10	10	12	16
	9	13	11	15	14
Σ	50	60	55	65	70

$$\begin{aligned} SS_T &= \frac{(50)^2}{5} + \frac{(60)^2}{5} + \frac{(55)^2}{5} + \frac{(65)^2}{5} + \frac{(70)^2}{5} - \frac{(300)^2}{25} \\ &= 3650 - 3600 \\ &= 50 \end{aligned}$$

and this sum of squares will have $k - 1$ d.f.

For each value of X we can now find the sum of squared deviations of the $n = 5$ values of Y from the mean value, $\overline{Y}_i$, associated with that value of X. We are primarily interested in the sum of these sums of squared deviations. This pooled sum of squares can be obtained by subtraction, as we have pointed out previously, and is commonly referred to as the within treatment sum of squares, or SS_W. Thus

$$SS_W = SS_{tot} - SS_T \tag{12.2}$$

and SS_W will have $k(n - 1)$ d.f. In our example, we have

$$SS_W = 100 - 50 = 50$$

with $5(5 - 1) = 20$ d.f. If we divide SS_W by its degrees of freedom, we obtain an unbiased estimate of the common population variance σ_Y^2 of the Y values associated with each value of X. This variance estimate is a mean square and is commonly referred to as the *mean square within treatments*. Thus

$$MS_W = \frac{SS_W}{k(n - 1)}$$

In our example, we have

$$MS_W = \frac{50}{5(5 - 1)} = 2.5$$

If the Y means associated with each value of X are in no way dependent on the particular values of X, that is, if $\mu_1 = \mu_2 = \cdots = \mu_k$, then it can be shown that the *treatment mean square* or

$$MS_T = \frac{SS_T}{k - 1}$$

is also an unbiased estimate of the same population variance as that estimated by MS_W. In our example, we have

$$MS_T = \frac{50}{5 - 1} = 12.5$$

and MS_T is considerably larger than MS_W.

To determine whether MS_T is significantly larger than MS_W, we have

$$F = \frac{MS_T}{MS_W}$$

with $k - 1$ d.f. for the numerator and $k(n - 1)$ d.f. for the denominator. In our example, we have

$$F = \frac{12.5}{2.5} = 5.0$$

a significant value with $\alpha = .01$ and with 4 and 20 d.f. It is reasonable to conclude that the Y means are not independent of the values of X.

12.3 Components of the Trend

Our primary interest, however, is the trend of the means. Figure 12.1 shows a plot of the Y means against the values of X. We know that an equation of the form

$$\overline{Y}_i' = \overline{Y} + b_1x_1 + b_2x_2 + b_3x_3 + b_4x_4 \tag{12.3}$$

where the values of x are the coefficients for orthogonal polynomials will predict perfectly the corresponding values of $\overline{Y}_i$. But some of the components of the trend represented by

$$\frac{[\Sigma(x_i \Sigma Y_i)]^2}{n\Sigma x_i^2}$$

may simply represent random variation and in this instance we may regard the values of

$$b_i = \frac{\Sigma(x_i \Sigma Y_i)}{n\Sigma x_i^2} \tag{12.4}$$

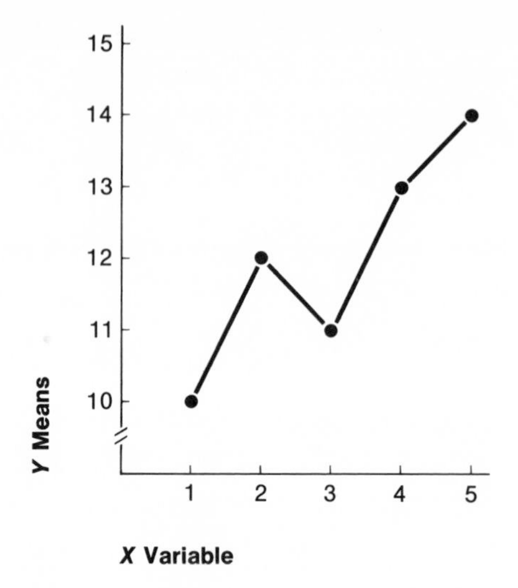

Figure 12.1 Plot of $k = 5$ means for equally spaced values of a quantitative variable X.

as not differing significantly from zero. Consequently, these terms in (12.3) may be dropped. We shall proceed systematically to investigate the components of the trend, starting with the linear component.

12.4 The Linear Component and a Test of Significance

To calculate the linear component of the trend, we use the coefficients x_1 for $k = 5$ as obtained from Table X in the Appendix. Then, we have

$$\frac{1}{n}\Sigma(x_1\Sigma Y_i) = \frac{1}{5}[(-2)(50) + (-1)(60) + (0)(55) + (1)(65) + (2)(70)]$$

$$= \frac{1}{5}(45)$$

and

$$b_1 = \frac{\Sigma(x_1\Sigma Y_i)}{n\Sigma x_1^2} = \frac{45}{(5)(10)} = .90$$

For the sum of squares for the linear component of the trend, we have

$$SS_L = \frac{[\Sigma(x_1\Sigma Y_1)]^2}{n\Sigma x_1^2} = \frac{(45)^2}{(5)(10)} = 40.5$$

with 1 d.f. Because SS_L has only 1 d.f., the mean square for the linear component will be equal to SS_L, that is, $MS_L = SS_L/1 = SS_L$.[1] To determine whether the linear component is significant, we find

$$F = \frac{MS_L}{MS_W} = \frac{SS_L}{MS_W} = \frac{40.5}{2.5} = 16.2$$

with 1 and 20 d.f. Because $F = 16.2$ is highly significant, we conclude that there is a significant linear component in the trend of the means. Our regression equation will then be

$$\overline{Y}_i' = \overline{Y} + b_1 x_1$$

where $\overline{Y} = 12$ and $b_1 = .90$ and the values of x_1 are the coefficients for the linear component.

In our example, we have

$$\begin{aligned} \overline{Y}_1' &= 12 + (.90)(-2) = 10.20 \\ \overline{Y}_2' &= 12 + (.90)(-1) = 11.10 \\ \overline{Y}_3' &= 12 + (.90)(0) = 12.00 \\ \overline{Y}_4' &= 12 + (.90)(1) = 12.90 \\ \overline{Y}_5' &= 12 + (.90)(2) = 13.80 \end{aligned}$$

The residual sum of squares will then be given by

$$SS_{res} = n\Sigma(\overline{Y}_i - \overline{Y}_i')^2 \tag{12.5}$$

where $\overline{Y}_i' = \overline{Y} + b_1 x_1$. In our example, we have

$$\begin{aligned} SS_{res} &= 5[(10 - 10.2)^2 + (12 - 11.1)^2 + (11 - 12.0)^2 \\ &\quad + (13 - 12.9)^2 + (14 - 13.8)^2] \\ &= 9.5 \end{aligned}$$

Ordinarily we would not use (12.5) to calculate SS_{res} because this sum of squares can be obtained much more easily by subtraction.[2] Thus

$$SS_{res} = SS_T - SS_L \tag{12.6}$$

In our example, we have $SS_T = 50.0$ and $SS_L = 40.5$ and, consequently,

$$SS_{res} = 50.0 - 40.5 = 9.5$$

which is equal to the value we obtained by direct calculation using (12.5).

---

[1] In the analysis of variance, as we have pointed out previously, sums of squares divided by their degrees of freedom are commonly referred to as mean squares. When a sum of squares, such as SS_L, has only 1 d.f., the sum of squares is itself a mean square.

[2] The proof of (12.6) is given in the answer to one of the exercises at the end of this chapter.

SS_{res} measures the deviations of the Y means from linearity and will have $k - 2$ d.f. To determine whether $MS_{res} = SS_{res}/(k - 2)$ is significant, we find

$$F = \frac{MS_{res}}{MS_W} \tag{12.7}$$

and, in our example,

$$F = \frac{9.5/(5 - 2)}{2.5} = 1.27$$

with 3 and 20 d.f., and this is a nonsignificant value of F with $\alpha = .05$.

Because there is a significant linear component of the trend and a nonsignificant residual variance, we would customarily end the investigation at this point. For purposes of illustration, however, we proceed to calculate the quadratic, cubic, and quartic components of the trend.

12.5 The Quadratic Component and a Test of Significance

Using the coefficients x_2 for the quadratic component, we obtain

$$\begin{aligned}\frac{1}{n}\Sigma(x_2\Sigma Y_i) &= \tfrac{1}{5}[(2)(50) + (-1)(60) + (-2)(55) + (-1)(65) \\ &\quad + (2)(70)] \\ &= \tfrac{1}{5}(5)\end{aligned}$$

Then

$$b_2 = \frac{\Sigma(x_2\Sigma Y_i)}{n\Sigma x_2^2} = \frac{5}{(5)(14)} = .07$$

and

$$SS_Q = \frac{[\Sigma(x_2\Sigma Y_i)]^2}{n\Sigma x_2^2} = \frac{(5)^2}{(5)(14)} = .36$$

with 1 d.f.

It is obvious that because

$$F = \frac{MS_Q}{MS_W} = \frac{.36}{2.50} < 1.00$$

the quadratic component is not significant. However, if we were to include the quadratic term in the regression equation, we would now have

$$\overline{Y}_i' = \overline{Y} + b_1x_1 + b_2x_2$$

or

$$\overline{Y}_i' = 12 + .90x_1 + .07x_2$$

and the resulting values of $\overline{Y}_i'$ would be

$$\overline{Y}_1' = 10.34$$
$$\overline{Y}_2' = 11.03$$
$$\overline{Y}_3' = 11.86$$
$$\overline{Y}_4' = 12.83$$
$$\overline{Y}_5' = 13.94$$

The new residual sum of squares,

$$SS_{res} = n\Sigma(\overline{Y}_i - \overline{Y}_i')^2$$

where $\overline{Y}_i'$ is now equal to $\overline{Y} + b_1x_1 + b_2x_2$, could be calculated directly and would be found to be equal to

$$SS_{res} = SS_T - SS_L - SS_Q \quad (12.8)$$

or

$$SS_{res} = 50.00 - 40.50 - .36 = 9.14$$

with $k - 3 = 2$ d.f.

It is obvious that taking the quadratic term into account in the regression equation has not reduced the residual sum of squares by any considerable amount.

12.6 The Cubic Component and a Test of Significance

Using the coefficients x_3 for the cubic component, we obtain

$$\frac{1}{n}\Sigma(x_3\Sigma Y_i) = \frac{1}{5}[(-1)(50) + (2)(60) + (0)(55) + (-2)(65) + (1)(70)]$$
$$= \frac{1}{5}(10)$$

Then

$$b_3 = \frac{\Sigma(x_3\Sigma Y_i)}{n\Sigma x_3^2} = \frac{10}{(5)(10)} = .20$$

and the sum of squares for the cubic component will be

$$SS_C = \frac{[\Sigma(x_3\Sigma Y_i)]^2}{n\Sigma x_3^2} = \frac{(10)^2}{(5)(10)} = 2.0$$

with 1 d.f.

A test of significance of the cubic component will be given by

$$F = \frac{MS_C}{MS_W}$$

In our example, $MS_C = 2.0$ and $MS_W = 2.5$; F is less than 1.00 and is nonsignificant.

Taking into account the linear, quadratic, and cubic terms of (12.3), we have

$$\overline{Y}_i' = \overline{Y} + b_1x_1 + b_2x_2 + b_3x_3$$

or

$$\overline{Y}_i' = 12 + .90x_1 + .07x_2 + .20x_3$$

and the predicted means will now be

$$\overline{Y}_1' = 10.14$$
$$\overline{Y}_2' = 11.43$$
$$\overline{Y}_3' = 11.86$$
$$\overline{Y}_4' = 12.43$$
$$\overline{Y}_5' = 14.14$$

If we were now to calculate the new residual sum of squares, using $\overline{Y}_i' = 12 + .90x_1 + .07x_2 + .20x_3$ as the predicted value of the Y means, we would find that

$$SS_{res} = n\Sigma(\overline{Y}_i - \overline{Y}_i')^2 = SS_T - SS_L - SS_Q - SS_C \quad (12.9)$$

or, in our example,

$$SS_{res} = 50.00 - 40.50 - .36 - 2.00 = 7.14$$

with $k - 4 = 1$ d.f.

12.7 The Quartic Component and a Test of Significance

The sum of squares for the quartic component must be equal to 7.14 because we know that SS_T is equal to the sum of the linear, quadratic, cubic, and quartic components. To show that this is so, we use the coefficients x_4 for the quartic component to obtain

$$\frac{1}{n}\Sigma(x_4\Sigma Y_i) = \frac{1}{5}[(1)(50) + (-4)(60) + (6)(55) + (-4)(65) + (1)(70)]$$

$$= \frac{1}{5}(-50)$$

and

$$b_4 = \frac{\Sigma(x_4 \Sigma Y_i)}{n\Sigma x_4^2} = \frac{-50}{(5)(70)} = -.14$$

Then the sum of squares for the quartic component will be

$$\frac{[\Sigma(x_4 \Sigma Y_i)]^2}{n\Sigma x_4^2} = \frac{(-50)^2}{(5)(70)} = 7.14$$

with 1 d.f.

As a test of significance of the quartic component, we have

$$F = \frac{7.14}{2.50} = 2.86$$

with 1 and 20 d.f., a nonsignificant value with $\alpha = .05$.

Taking into account the linear, quadratic, cubic, and quartic terms in (12.3), we now have

$$\overline{Y}_i' = \overline{Y} + b_1x_1 + b_2x_2 + b_3x_3 + b_4x_4$$

or

$$\overline{Y}_i' = 12 + .90x_1 + .07x_2 + .20x_3 + .14x_4$$

and the predicted means are now

$$\overline{Y}_1' = 10.00$$
$$\overline{Y}_2' = 11.99$$
$$\overline{Y}_3' = 11.02$$
$$\overline{Y}_4' = 12.99$$
$$\overline{Y}_5' = 14.00$$

These means are equal, within rounding errors, to the observed values of the means.

12.8 Correlations of $\overline{Y}_i$ with the Orthogonal Coefficients

We now note that the sum of the squares of the correlation coefficients

$$r^2_{x_1\overline{Y}_i} = \frac{[\Sigma(x_1 \Sigma Y_i)]^2}{(n\Sigma x_1^2)(SS_T)} = \frac{(45)^2}{(5)(10)(50)} = .810$$

$$r^2_{x_2\overline{Y}_i} = \frac{[\Sigma(x_2 \Sigma Y_i)]^2}{(n\Sigma x_2^2)(SS_T)} = \frac{(-5)^2}{(5)(14)(50)} = .007$$

$$r^2_{x_3\overline{Y}_i} = \frac{[\Sigma(x_3 \Sigma Y_i)]^2}{(n\Sigma x_3^2)(SS_T)} = \frac{(10)^2}{(5)(10)(50)} = .040$$

$$r^2_{x_4\overline{Y}_i} = \frac{[\Sigma(x_4 \Sigma Y_i)]^2}{(n\Sigma x_4^2)(SS_T)} = \frac{(-50)^2}{(5)(70)(50)} = .143$$

is equal to 1.00. In this example, each of the sets of orthogonal coefficients accounts for a proportion of the treatment sum of squares (SS_T). The largest proportion (.81) is accounted for by the coefficients for the linear component.

12.9 Multiple Correlation Coefficient

The correlation coefficient between the actual Y means and the predicted values

$$\overline{Y}_i' = \overline{Y} + b_1x_1 + b_2x_2 + b_3x_3 + b_4x_4$$

is known as a multiple correlation coefficient and is represented by R. In our example,

$$R_{\overline{Y}_i\overline{Y}_i'} = R_{\overline{Y}_i.x_1x_2x_3x_4}$$

is the multiple correlation coefficient between $\overline{Y}_i$ and a weighted linear sum of x_1, x_2, x_3, and x_4, the weights being the corresponding regression coefficients b_1, b_2, b_3, and b_4. Because the values of x_1, x_2, x_3, and x_4 are orthogonal or uncorrelated, the square of the multiple correlation coefficient will be

$$R^2_{\overline{Y}_i\overline{Y}_i'} = r^2_{x_1\overline{Y}_i} + r^2_{x_2\overline{Y}_i} + r^2_{x_3\overline{Y}_i} + r^2_{x_4\overline{Y}_i}$$

or, in our example,

$$R^2_{\overline{Y}_i\overline{Y}_i'} = .810 + .007 + .040 + .143 = 1.000$$

The square of the multiple correlation coefficient will always be equal to 1.00 if all $k - 1$ terms are used in the regression equation. If some of the b values are equal to zero so that the corresponding values of r are equal to zero, then $R^2_{\overline{Y}_i\overline{Y}_i'}$ will still be equal to 1.00, if all of the remaining terms in the regression equation are used. It is, of course, possible for $R^2_{\overline{Y}_i\overline{Y}_i'}$ to be equal to 1.00 when only one term is included in the regression equation. That was true of the examples cited in the preceding chapter.

The square of the multiple correlation coefficient is simply the proportion of SS_T that can be accounted for by a weighted sum of the x values. For example, using only the x_1 and x_4 values, we have

$$Y' = \overline{Y} + b_1x_1 + b_4x_4$$

and

$$R^2_{\overline{Y}_i\overline{Y}_i'} = .810 + .143 = .953$$

Approximately 95 percent of the treatment sum of squares can be accounted for by the linear and quartic components.

Exercises

12.1. We have $k = 5$ equally spaced values of an independent variable X and $n = 5$ observations of a dependent variable for each value of X, as shown in the following table. Make all tests of significance with $\alpha = .05$.

X_1	X_2	X_3	X_4	X_5
3	7	8	4	10
5	9	10	6	9
2	8	6	5	11
1	6	7	8	8
4	5	9	7	7

(a) Calculate SS_T and SS_W and determine whether MS_T is significant. (b) What are the values of b_1, b_2, b_3, and b_4? (c) Calculate the linear, quadratic, cubic, and quartic components of SS_T and test each for significance. (d) Are the deviations from linearity significant? (e) Are the deviations from $\overline{Y}_i' = \overline{Y} + b_1x_1 + b_2x_2$ significant? (f) Are the deviations from $\overline{Y}_i' = \overline{Y} + b_1x_1 + b_2x_2 + b_3x_3$ significant? (g) Calculate the squares of the correlation coefficients between $\overline{Y}_i$ and each of the sets of orthogonal coefficients. (h) What proportion of SS_T can be accounted for by the linear and cubic components? (i) Calculate the values of $\overline{Y}_i' = \overline{Y} + b_1x_1 + b_2x_2 + b_3x_3 + b_4x_4$.

12.2. Given that $\overline{Y}_i' = \overline{Y} + b_1x_1$, prove that the residual sum of squares

$$SS_{res} = n\Sigma(\overline{Y}_i - \overline{Y}_i')^2$$

is equal to $SS_T - SS_L$.

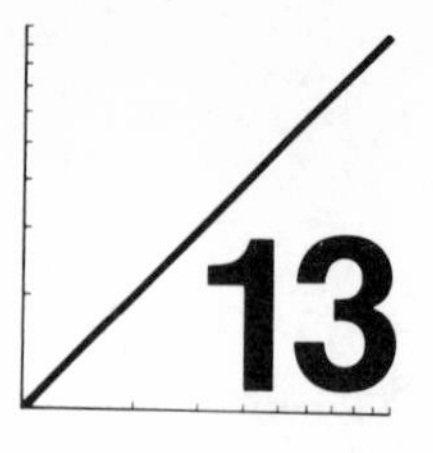

13 Analysis of Variance for a Simple Repeated Measure Design

13.1 Introduction

In the experiment described in the preceding chapter, a different group of subjects was assigned at random to each of the values of the quantitative independent variable (X) so that for each subject only one value of the dependent variable (Y) was obtained. Experiments of this kind are often described as involving a *completely randomized design* or a *randomized group design*. In contrast to a completely randomized design, we may have an experiment involving a *repeated measure design*. With a repeated measure design, a single group of subjects is tested with each value of X so that we have a Y value for each of the subjects for each of the values of X.

Suppose, for example, that an experimenter is interested in the change in the blood sugar level (Y) of dogs when they are injected with differing amounts of insulin (X). If the same dogs receive each of the different injections of insulin, it would obviously be necessary to have some assurance that the effects of a given injection have been dissipated or worn off before the animal is given another injection. In other words, if we are interested in studying the relationship between blood sugar level and amount of insulin injected, using a repeated measure design, we would want to be sure that there are no *carry-over* effects from one dosage to another. One way to accomplish this would be to separate the different injections by a sufficient period so that any effect of a previous dosage would no longer be present. A completely randomized design, on the other hand, would, of course, eliminate any possibility of carry-over effects.

In some experiments the carry-over effects themselves are of interest. For example, we might be interested in the change in the number of correct responses (Y) made by subjects in learning a list of paired associates when the subjects are given a series of trials (X). In this experiment, we might expect an improvement in performance or a decrease in the number of errors from trial to trial as a result of practice or learning on previous trials, and the cumulative carry-over effects of the previous trials on a given trial would be of experimental interest.

In this chapter we illustrate by means of a simple example the analysis of variance for a repeated measure design. We again assume that the values of the X variable are equally spaced. In learning experiments in which the successive trials comprise the different values of X, the time intervals between trials should be equally spaced.

13.2 An Experiment Involving Four Repeated Measures

Assume that we have a maze consisting of ten choice points. If a rat is placed in the starting box and makes a correct turn at each choice point, the rat will end up in a goal box where food has been placed. Instead of observing the number of correct turns we observe the number of wrong turns or errors that the rat makes in arriving at the goal box. It is our belief that the mean number of errors should decrease linearly if rats are given a number of successive trials. The time intervals between the trials are equally spaced. Table 13.1 gives the number of errors made by each of five rats on each of four successive trials.

Figure 13.1 shows a plot of the mean number of errors for each of the four trials. The trend of the means is predominantly linear with some slight degree of curvature. We wish to determine whether the linear component of the trial sum of squares is significant and also whether the trial means deviate significantly from linearity.

TABLE 13.1 Number of errors made by five rats on each of four successive trials

	Trials					
Subjects	1	2	3	4	Σ	Means
1	8	8	7	5	28	7.00
2	7	7	5	4	23	5.75
3	9	4	4	4	21	5.25
4	6	4	3	3	16	4.00
5	5	3	2	2	12	3.00
Σ	35	26	21	18	100	

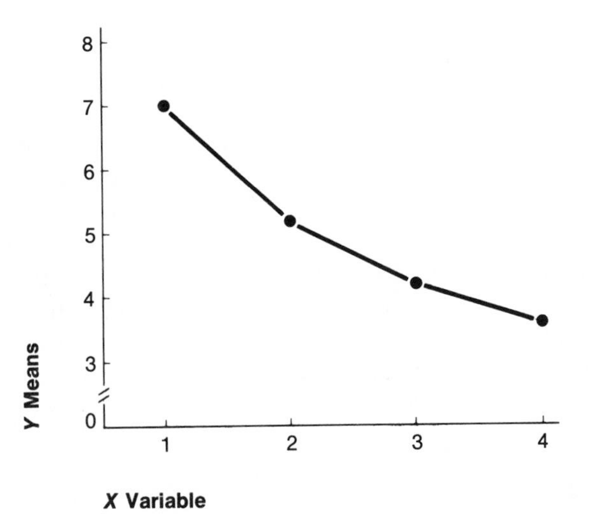

Figure 13.1 Plot of $k = 4$ means for equally spaced values of a quantitative variable X.

13.3 Analysis of Variance for the Repeated Measure Design

We begin the analysis of variance in the manner described in the preceding chapter. For the total sum of squares, we have

$$SS_{tot} = (8)^2 + (7)^2 + \cdots + (2)^2 - \frac{(100)^2}{20}$$
$$= 582.0 - 500.0$$
$$= 82.0$$

with $kn - 1 = 19$ d.f. For the trial or treatment sum of squares, we have

$$SS_T = \frac{(35)^2}{5} + \frac{(26)^2}{5} + \frac{(21)^2}{5} + \frac{(18)^2}{5} - \frac{(100)^2}{20}$$
$$= 533.2 - 500.0$$
$$= 33.2$$

with $k - 1 = 3$ d.f.

Then the within treatment sum of squares can be either calculated directly or obtained by subtraction. By subtraction, we have

$$SS_W = SS_{tot} - SS_T = 82.0 - 33.2 = 48.8$$

with $k(n - 1) = 16$ d.f. If the experiment involved a completely randomized design and different subjects were tested with each value of X, then

$$MS_W = \frac{48.8}{16} = 3.05$$

would be the appropriate error mean square for testing the linear, quadratic, and cubic components of the trial sum of squares for significance.

With a repeated measure design, however, the pooled within treatment sum of squares consists of two components. One of these components is called the *subject sum of squares* (SS_S) and represents the variation in the performance of the different subjects averaged over the k trials. The other component represents a residual sum of squares and is called the *subject* $\times$ *treatment* or *subject* $\times$ *trial* sum of squares (SS_{ST}).

The subject sum of squares (SS_S) is obtained by first adding the values of Y for each subject. These sums are shown at the right in Table 13.1. The subject sum of squares is a measure of the variation of the means of the subjects about the overall mean, just as the trial sum of squares is a measure of the variation of the trial means about the overall mean. The calculation of the subject sum of squares is analogous to the calculation of the trial sum of squares. In Table 13.1 the means for the five subjects are 7.00, 5.75, 5.25, 4.00, and 3.00, respectively, and the overall mean is equal to 5.00. Then

$$\begin{aligned} SS_S &= 4[(7.00 - 5.00)^2 + (5.75 - 5.00)^2 + \cdots + (3.00 - 5.00)^2] \\ &= (4)(9.625) \\ &= 38.5 \end{aligned}$$

or, equivalently,

$$\begin{aligned} SS_S &= \frac{(28)^2}{4} + \frac{(23)^2}{4} + \frac{(21)^2}{4} + \frac{(16)^2}{4} + \frac{(12)^2}{4} - \frac{(100)^2}{20} \\ &= 538.5 - 500.0 \\ &= 38.5 \end{aligned}$$

with $n - 1 = 4$ d.f.

If we now subtract the subject sum of squares from the within treatment or within trial sum of squares, we obtain the subject $\times$ trial sum of squares with $(n - 1)(k - 1)$ d.f., or

$$SS_{ST} = SS_W - SS_S \tag{13.1}$$

and, in our example,

$$SS_{ST} = 48.8 - 38.5 = 10.3$$

with $(5 - 1)(4 - 1) = 12$ d.f. Dividing the subject $\times$ trial sum of squares by its degrees of freedom, we obtain the mean square

$$MS_{ST} = \frac{10.3}{12} = .86$$

and MS_{ST} is the appropriate error mean square for the tests of significance of interest. Note that $MS_{ST} = .86$ is considerably smaller than $MS_W = 3.05$. We shall show why this is so later in this chapter.

13.4 Trend Components of the Trial Sum of Squares

The linear component of the trial sum of squares is obtained in the same manner as described in the preceding chapter. With $k = 4$ values of X, the coefficients x_1 for the linear component, as obtained from Table X, are -3, -1, 1, and 3. Then

$$\frac{1}{n}\Sigma(x_1 \Sigma Y_i) = \frac{1}{5}[(-3)(35) + (-1)(26) + (1)(21) + (3)(18)]$$

$$= \frac{1}{5}(-56)$$

and

$$b_1 = \frac{\Sigma(x_1 \Sigma Y_i)}{n\Sigma x_1^2} = \frac{-56}{(5)(20)} = -.56$$

For the sum of squares for the linear component we have

$$SS_L = \frac{[\Sigma(x_1 \Sigma Y_i)]^2}{n\Sigma x_1^2} = \frac{(-56)^2}{(5)(20)} = 31.36$$

with 1 d.f. To determine whether the linear component is significant, we find

$$F = \frac{MS_L}{MS_{ST}} = \frac{31.36}{.86} = 36.47$$

and this is a highly significant value of F with 1 and 12 d.f.

To determine whether the trial means deviate significantly from linearity, we find the sum of squares for deviations from linearity, or

$$SS_{res} = SS_T - SS_L$$

and, in our example,

$$SS_{res} = 33.20 - 31.36 = 1.84$$

with $k - 2 = 2$ d.f. Then $MS_{res} = 1.84/2 = .92$, and

$$F = \frac{MS_{res}}{MS_{ST}} = \frac{.92}{.86} = 1.07$$

This is a nonsignificant value of F with 2 and 12 d.f.

We see that our major hypotheses are confirmed. The linear component of the trial sum of squares is significant and the deviations from linearity are nonsignificant. Ordinarily, our analysis would stop at this point, but for purposes of illustration we also calculate the quadratic and cubic components of the trial sum of squares.

For the quadratic component the values of the coefficients x_2, as obtained from Table X, are 1, -1, -1, and 1, respectively. Then

$$\frac{1}{n}\Sigma(x_2\Sigma Y_i) = \frac{1}{5}[(1)(35) + (-1)(26) + (-1)(21) + (1)(18)]$$

$$= \frac{1}{5}(6)$$

and

$$b_2 = \frac{\Sigma(x_2\Sigma Y_i)}{n\Sigma x_2^2} = \frac{(6)}{(5)(4)} = .30$$

and

$$SS_Q = \frac{[\Sigma(x_2\Sigma Y_i)]^2}{n\Sigma x_2^2} = \frac{(6)^2}{(5)(4)} = 1.80$$

Similarly, using the coefficients -1, 3, -3, and 1, as obtained from Table X, for the cubic component, we have

$$\frac{1}{n}\Sigma(x_3\Sigma Y_i) = \frac{1}{5}[(-1)(35) + (3)(26) + (-3)(21) + (1)(18)]$$

$$= \frac{1}{5}(-2)$$

and

$$b_3 = \frac{\Sigma(x_3\Sigma Y_i)}{n\Sigma x_3^2} = \frac{-2}{(5)(20)} = -.02$$

and

$$SS_C = \frac{[\Sigma(x_3\Sigma Y_i)]^2}{n\Sigma x_3^2} = \frac{(-2)^2}{(5)(20)} = .04$$

We also note that

$$\begin{aligned} SS_T &= SS_L + SS_Q + SS_C \\ &= 31.36 + 1.80 + .04 \\ &= 33.20 \end{aligned}$$

Obviously, the linear component accounts for almost all of the trial sum of squares, and the quadratic and cubic components are negligible. We may, therefore, assume that the linear equation

$$\overline{Y}_i' = \overline{Y} + b_1 x_1$$

or

$$\overline{Y}_i' = 5.00 - .56x_1$$

describes the trend of the trial means within the limits of random sampling.

13.5 The Relationship between MS_{ST} and MS_W

Using a repeated measure design for the experiment described, we have partitioned the total sum of squares into the three major components to the right of the equals sign in the following equation:

$$SS_{tot} = SS_T + SS_S + SS_{ST} \tag{13.2}$$

Then we also have

$$SS_{ST} = SS_{tot} - SS_T - SS_S \tag{13.3}$$

But, as we have shown previously, $SS_{tot} - SS_T = SS_W$ and, consequently,

$$SS_{ST} = SS_W - SS_S \tag{13.4}$$

Because SS_S can never be negative, we see that SS_{ST} can never be larger than SS_W.

It can also be shown[1] that

$$MS_{ST} = MS_W - \bar{c}_{ij}$$

where $\bar{c}_{ij}$ is the average value of the covariances between the values of Y for the various trials. For example, Table 13.2 gives the values of $Y_i - \overline{Y}_i$, that is, the deviations of the values of Y for each trial from the trial mean. Then the covariance of the Y values for Trials i and j will be given by

$$c_{ij} = \frac{1}{n-1}\Sigma(Y_i - \overline{Y}_i)(Y_j - \overline{Y}_j)$$

From the values given in Table 13.2, we see that the covariance between the Y values for Trials 1 and 2 is

$$c_{12} = \frac{1}{5-1}[(1.0)(2.8) + (0.0)(1.8) + \cdots + (-2.0)(-2.2)]$$

$$= \frac{1}{4}(6)$$

$$= 1.50$$

[1]A proof can be found in A. L. Edwards, *Statistical Methods* (4th ed.). New York: Holt, Rinehart and Winston, 1973.

TABLE 13.2 Values of $Y_i - \overline{Y}_i$ for the data of Table 13.1

Subjects	Trials 1	2	3	4
1	1.0	2.8	2.8	1.4
2	0.0	1.8	0.8	0.4
3	2.0	−1.2	−0.2	0.4
4	−1.0	−1.2	−1.2	−0.6
5	−2.0	−2.2	−2.2	−1.6
Σ	0.0	0.0	0.0	0.0

Similar calculations would show that $c_{13} = 2.00$, $c_{14} = 1.50$, $c_{23} = 3.95$, $c_{24} = 2.10$, and $c_{34} = 2.10$. Then

$$\bar{c}_{ij} = \frac{1.50 + 2.00 + 1.50 + 3.95 + 2.10 + 2.10}{6} = 2.19$$

and, consequently,

$$MS_{ST} = MS_W - \bar{c}_{ij} = 3.05 - 2.19 = .86$$

Only if $\bar{c}_{ij}$ is positive will MS_{ST} be smaller than MS_W.

In general, if there are individual differences in the subjects tested in the experiment such that some subjects tend to have Y values that are consistently above the mean Y value for each value of X and other subjects tend to have Y values that are consistently below the mean Y value for each value of X, then $\bar{c}_{ij}$ will be positive. For example, we observe in Table 13.2 that for the first two subjects the values of $Y_i - \overline{Y}_i$ are positive for all values of X. For the last two subjects the values of $Y_i - \overline{Y}_i$ are negative for all values of X. When all of the covariances are positive, the subject sum of squares will tend to be large and, consequently, $SS_{ST} = SS_W - SS_S$ will tend to be small.

The major advantage of a repeated measure design, when it is appropriate to use this design, is that MS_{ST} will be smaller than MS_W, the error mean square for a completely randomized design. This, in turn, means that the tests of significance of the linear, quadratic, and other components of the treatment sum of squares will be more sensitive with a repeated measure design than with a completely randomized design.

Exercises

13.1. Assume that we have a completely randomized design with three equally spaced values of X and that five subjects are assigned at random to each value of X. The Y values for each subject are as follows:

	X_1	X_2	X_3
	3	7	6
	1	4	2
	2	3	4
	4	6	3
	5	5	5
Σ	15	25	20

(a) Find the values of MS_W and MS_T and determine whether the means differ significantly. (b) Find the linear and quadratic components of the treatment sum of squares. (c) Will the quadratic component be equal to the sum of squares for deviations from linearity? Explain why or why not. (d) Is the linear component significant? (e) Are there significant deviations from linearity?

13.2. Now assume that the data given in exercise 13.1 represent a repeated measure design and that the same five subjects are tested with each of the three different values of X. (a) Find the subject sum of squares and the subject $\times$ treatment sum of squares. (b) Is MS_T significant with the repeated measure design? (c) Is the linear component of the treatment sum of squares significant? (d) Are the deviations from linearity significant? (e) Subtract the treatment means from the corresponding values of Y. Calculate the covariances: c_{12}, c_{13}, and c_{23}. (f) What is the average value of the covariances? (g) Is MS_{ST} equal to $MS_W - \bar{c}_{ij}$?

13.3. (a) Using the data of Table 13.1, find the values of $\overline{Y}_i' = \overline{Y} + b_1x_1$. (b) Find the correlation coefficient between $\overline{Y}_i$ and $\overline{Y}_i'$. (c) What proportion of the treatment sum of squares can be accounted for by the linear component?

13.4. (a) Using the data of Table 13.1, find the values of $\overline{Y}_i' = \overline{Y} + b_1x_1 + b_2x_2$. (b) Find the correlation coefficient between $\overline{Y}_i$ and $\overline{Y}_i'$. (c) What proportion of the treatment sum of squares can be accounted for by the combined linear and quadratic components?

13.5. (a) Using the data of Table 13.1, find the values of $\overline{Y}_i' = \overline{Y} + b_1x_1 + b_2x_2 + b_3x_3$. (b) Find the correlation coefficient between $\overline{Y}_i$ and $\overline{Y}_i'$. (c) What proportion of the treatment sum of squares can be accounted for by the combined linear, quadratic, and cubic components?

13.6. Subjects learn a list of twenty-five paired associates. The experimenter plans to test the subjects for recall of the paired associates 1, 2, 3, 4, and 5 days after learning. Consider the following two designs: (1) The subjects are divided at random into five groups of ten subjects each. A different group of subjects is tested on each of the five days after learning. (2) A single group of ten subjects is tested on each of the five days after learning. (a) How many degrees of freedom will the error mean squares have in each of the two designs? (b) Would you expect the trend of the means in the two designs to be the same? Explain why or why not.

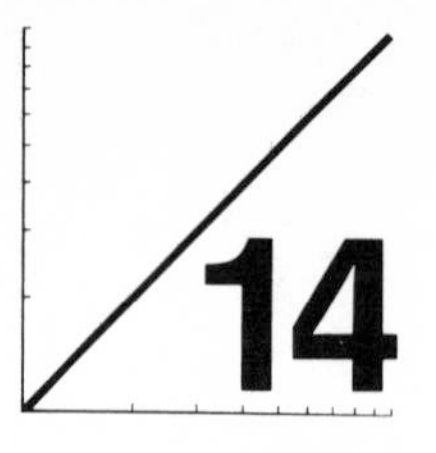

Multiple Correlation and Regression

14.1 Introduction

Suppose that for a random sample of n subjects we have available for each subject measures of three or more variables. One of the variables is of primary interest in that we would like to predict it from some weighted linear combination of the other variables. We shall refer to the variable we are interested in predicting as the Y variable and the remaining k variables as the X variables. Thus, we have a multiple regression equation

$$Y' = a + b_1X_1 + b_2X_2 + \cdots + b_kX_k \tag{14.1}$$

and we want to find the values of $a, b_1, b_2, \ldots, b_k$ that will result in the highest possible positive correlation between the observed values of Y and the predicted values or Y'. If this is done, then the resulting correlation coefficient between Y and Y' is called a *multiple correlation coefficient*[1] and is represented by $R_{YY'}$.

As the number of X variables increases, the calculations involved in finding the values of $b_1, b_2, \ldots, b_k$, become complex and overwhelming,[2] although they can be accomplished quickly and easily by a high-speed electronic computer. Fortunately, many of the principles of multiple regression and correlation can be illustrated with an example consisting of one Y variable and two X variables. The calculations used to solve the three-variable problem are relatively simple.

---

[1]The notation $R_{Y.123\ldots k}$ is also used for the multiple correlation coefficient.

[2]This statement is correct when the X variables are intercorrelated, as they typically are. The calculations are relatively simple when the X variables are mutually orthogonal, as we shall see later.

14.2 Calculating the Values of b_1 and b_2 for a Three-Variable Problem

With two X variables, the multiple regression equation becomes

$$Y' = a + b_1X_1 + b_2X_2 \tag{14.2}$$

The maximum value of the correlation coefficient between the observed Y values and the predicted Y' values will be obtained if we find the values of a, b_1, and b_2 that minimize the sum of squared errors of prediction or the residual sum of squares

$$SS_{res} = \Sigma(Y - Y')^2$$

The required value of a is given by the equation

$$a = \overline{Y} - b_1\overline{X}_1 - b_2\overline{X}_2 \tag{14.3}$$

Substituting this value in (14.2), we obtain

$$Y' = \overline{Y} + b_1(X_1 - \overline{X}_1) + b_2(X_2 - \overline{X}_2)$$

or

$$Y' = \overline{Y} + b_1x_1 + b_2x_2$$

Then, if the residual sum of squares is to be minimized, the values of b_1 and b_2 must satisfy the following equations:

$$b_1\Sigma x_1^2 + b_2\Sigma x_1x_2 = \Sigma x_1y \tag{14.4}$$

and

$$b_1\Sigma x_1x_2 + b_2\Sigma x_2^2 = \Sigma x_2y \tag{14.5}$$

We have two equations with two unknowns, b_1 and b_2, and these equations can be solved by using standard algebraic methods. For example, if we multiply (14.4) by Σx_2^2 and (14.5) by Σx_1x_2, we obtain

$$b_1(\Sigma x_1^2)(\Sigma x_2^2) + b_2(\Sigma x_1x_2)(\Sigma x_2^2) = (\Sigma x_1y)(\Sigma x_2^2) \tag{14.6}$$

and

$$b_1(\Sigma x_1x_2)^2 + b_2(\Sigma x_1x_2)(\Sigma x_2^2) = (\Sigma x_2y)(\Sigma x_1x_2) \tag{14.7}$$

Then, subtracting (14.7) from (14.6), we have

$$b_1(\Sigma x_1^2)(\Sigma x_2^2) - b_1(\Sigma x_1x_2)^2 = (\Sigma x_1y)(\Sigma x_2^2) - (\Sigma x_2y)(\Sigma x_1x_2)$$

or

$$b_1 = \frac{(\Sigma x_1y)(\Sigma x_2^2) - (\Sigma x_2y)(\Sigma x_1x_2)}{(\Sigma x_1^2)(\Sigma x_2^2) - (\Sigma x_1x_2)^2} \tag{14.8}$$

TABLE 14.1 Sums of squared deviations and sums of products of paired deviations for three variables, X_1, X_2, and Y, for a sample of $n = 20$ observations

Variable	X_1	X_2	Y	Means
X_1	76.0	45.6	212.8	4.0
X_2		304.0	364.8	8.0
Y			1216.0	20.0
s	2.0	4.0	8.0	

Following a similar procedure, we find that

$$b_2 = \frac{(\Sigma x_2 y)(\Sigma x_1^2) - (\Sigma x_1 y)(\Sigma x_1 x_2)}{(\Sigma x_1^2)(\Sigma x_2^2) - (\Sigma x_1 x_2)^2} \tag{14.9}$$

14.3 A Numerical Example of a Three-Variable Problem

Assume that we have a random sample of $n = 20$ with one Y variable and two X variables.[3] Table 14.1 gives the values of

$$\Sigma x_1^2 = \Sigma(X_1 - \overline{X}_1)^2 = 76.0$$
$$\Sigma x_2^2 = \Sigma(X_2 - \overline{X}_2)^2 = 304.0$$
$$\Sigma y^2 = \Sigma(Y - \overline{Y})^2 = 1216.0$$
$$\Sigma x_1 x_2 = \Sigma(X_1 - \overline{X}_1)(X_2 - \overline{X}_2) = 45.6$$
$$\Sigma x_1 y = \Sigma(X_1 - \overline{X}_1)(Y - \overline{Y}) = 212.8$$
$$\Sigma x_2 y = \Sigma(X_2 - \overline{X}_2)(Y - \overline{Y}) = 364.8$$

for the sample of $n = 20$ observations. The means of $\overline{X}_1$, $\overline{X}_2$, and $\overline{Y}$ are shown at the right of the table and the standard deviations appear at the bottom of the table. The diagonal entries in the table are the sums of squared deviations from the means. For example, we have $\Sigma x_1^2 = 76.0$. The nondiagonal entries give the sums of the products of the paired deviations. For example, we have $\Sigma x_1 x_2 = 45.6$. Substituting the appropriate values from Table 14.1 in (14.8) and (14.9), we have

[3]In general, the number of observations n in a multiple correlation problem should be considerably larger than the number of X variables k. No hard and fast rules are available, but we would suggest that n/k should be equal to or greater than 10.

$$b_1 = \frac{(212.8)(304.0) - (364.8)(45.6)}{(76.0)(304.0) - (45.6)^2} = 2.2857$$

and

$$b_2 = \frac{(364.8)(76.0) - (212.8)(45.6)}{(76.0)(304.0) - (45.6)^2} = .8571$$

Then, substituting in (14.3), we obtain

$$a = 20.0 - (2.2857)(4.0) - (.8571)(8.0) = 4.0$$

and the multiple regression equation will be

$$Y' = 4.0 + 2.2857X_1 + .8571X_2$$

14.4 Partitioning the Total Sum of Squares into the Regression and Residual Sums of Squares

We have previously shown that when there are two variables, X and Y, the total sum of squares for the Y variable or

$$SS_{tot} = \Sigma y^2 = \Sigma(Y - \overline{Y})^2$$

can be partitioned into two parts, the sum of squares for linear regression or

$$SS_{reg} = b\Sigma xy = \frac{(\Sigma xy)^2}{\Sigma x^2}$$

and the residual sum of squares

$$SS_{res} = \Sigma(Y - Y')^2$$

This can also be done with a multiple regression equation, that is,

$$SS_{tot} = SS_{reg} + SS_{res} \tag{14.10}$$

In a multiple regression problem, the sum of squares for linear regression will be given by

$$SS_{reg} = b_1\Sigma x_1 y + b_2\Sigma x_2 y + \cdots + b_k\Sigma x_k y \tag{14.11}$$

In our example, we have two values of b with $b_1 = 2.2857$ and $b_2 = .8571$. Substituting these two values, along with $\Sigma x_1 y = 212.8$ and $\Sigma x_2 y = 364.8$, as given in Table 14.1, in (14.11), we have

$$SS_{reg} = (2.2857)(212.8) + (.8571)(364.8) = 799.07$$

as the sum of squares for linear regression with $k = 2$ d.f. Then the residual sum of squares can be obtained by subtraction, that is,

$$SS_{res} = SS_{tot} - SS_{reg} \tag{14.12}$$

and, in our example, we have

$$SS_{res} = 1216.00 - 799.07 = 416.93$$

with $n - k - 1 = 17$ d.f.

14.5 The Multiple Correlation Coefficient

The square of the multiple correlation coefficient will then be given by

$$R^2_{YY'} = \frac{SS_{reg}}{SS_{tot}} \tag{14.13}$$

and, in our example, we have

$$R^2_{YY'} = \frac{799.07}{1216.00} = .6571$$

The square root of $R^2_{YY'}$ is, of course, the multiple correlation coefficient, and

$$R_{YY'} = \sqrt{.6571} = .8106$$

Note also that

$$1 - R^2_{YY'} = 1 - \frac{SS_{reg}}{SS_{tot}} = \frac{SS_{tot} - SS_{reg}}{SS_{tot}} = \frac{SS_{res}}{SS_{tot}} \tag{14.14}$$

$R^2_{YY'}$ is, in other words, the proportion of SS_{tot} that can be accounted for by linear regression, and $1 - R^2_{YY'}$ is the proportion of SS_{tot} that is independent of the linear regression of Y on Y'.

14.6 Calculating $R^2_{YY'}$ from the Correlation Coefficients

The correlation coefficients between X_1 and Y, X_2 and Y, and X_1 and X_2 can be obtained from the data given in Table 14.1. For example,

$$r_{1Y} = \frac{\Sigma x_1 y}{\sqrt{\Sigma x_1^2}\sqrt{\Sigma y^2}} = \frac{212.8}{\sqrt{76.0}\sqrt{1216.0}} = .70$$

$$r_{2Y} = \frac{\Sigma x_2 y}{\sqrt{\Sigma x_2^2}\sqrt{\Sigma y^2}} = \frac{364.8}{\sqrt{304.0}\sqrt{1216.0}} = .60$$

and

$$r_{12} = \frac{\Sigma x_1 x_2}{\sqrt{\Sigma x_1^2}\sqrt{\Sigma x_2^2}} = \frac{45.6}{\sqrt{76.0}\sqrt{304.0}} = .30$$

Then the multiple correlation coefficient squared will also be given by

$$R^2_{YY'} = \frac{r^2_{1Y} + r^2_{2Y} - 2r_{1Y}r_{2Y}r_{12}}{1 - r^2_{12}} \tag{14.15}$$

and, in our example, we have

$$R^2_{YY'} = \frac{(.70)^2 + (.60)^2 - 2(.70)(.60)(.30)}{1 - (.30)^2} = .6571$$

which is equal to the value we obtained before.

The regression coefficients b_1 and b_2 can also be obtained using the correlation coefficients. For example, if we divide both the numerator and denominator of (14.8) by $(n - 1)^2$ and simplify the resulting expression, we obtain

$$b_1 = \frac{s_Y(r_{1Y} - r_{2Y}r_{12})}{s_{X_1}(1 - r^2_{12})} \tag{14.16}$$

Similarly, dividing both the numerator and denominator of (14.9) by $(n - 1)^2$, we obtain

$$b_2 = \frac{s_Y(r_{2Y} - r_{1Y}r_{12})}{s_{X_2}(1 - r^2_{12})} \tag{14.17}$$

Substituting the appropriate values in (14.16) and (14.17), we have

$$b_1 = \frac{8[.70 - (.60)(.30)]}{2[1 - (.30)^2]} = 2.2857$$

and

$$b_2 = \frac{8[.60 - (.70)(.30)]}{4[1 - (.30)^2]} = .8571$$

which are equal to the values we obtained previously.

14.7 Tests of Significance for $R^2_{YY'}$

As (14.11) shows, if both b_1 and b_2 are equal to zero in a three-variable problem, then SS_{reg} will be equal to zero and $R^2_{YY'}$ will also be equal to zero. The obtained values of b_1 and b_2 are estimates of the corresponding population values β_1 and β_2. A test of the null hypothesis that the two population values β_1 and β_2 are *both* equal to zero will be given by

$$F = \frac{R^2_{YY'}/k}{(1 - R^2_{YY'})/(n - k - 1)} \tag{14.18}$$

or equivalently by

$$F = \frac{SS_{reg}/k}{SS_{res}/(n - k - 1)} = \frac{MS_{reg}}{MS_{res}} \tag{14.19}$$

For both F ratios the degrees of freedom for the numerator will be equal to k, the number of X variables, and the degrees of freedom for the denominator will be equal to $n - k - 1$.

Using (14.18), we have

$$F = \frac{.6571/2}{(1 - .6571)/17} = 16.29$$

or, equivalently, using (14.19), we have

$$F = \frac{799.07/2}{416.93/17} = 16.29$$

with 2 and 17 d.f. This is a significant value with $\alpha = .05$.

Unfortunately, the test of the null hypothesis $\beta_1 = \beta_2 = 0$ is not very satisfying. If the null hypothesis is rejected, then it may be because either β_1 or β_2 is not equal to zero or because both β_1 and β_2 are not equal to zero. There is a way, however, to test the null hypothesis $\beta_1 = 0$, given that X_2 has already been included in the regression equation, and also the null hypothesis $\beta_2 = 0$, given that X_1 has already been included in the regression equation. These conditional tests of significance are described in the next section.

14.8 Conditional Tests of Significance

When both X_1 and X_2 are included in the regression equation, the sum of squares for linear regression is equal to $SS_{reg} = 799.07$. Now suppose we find the regression sum of squares when only X_1 is included in the regression equation. Using the appropriate values given in Table 14.1, we have

$$SS_{reg} = \frac{(\Sigma x_1 y)^2}{\Sigma x_1^2} = \frac{(212.8)^2}{76.0} = 595.84$$

with 1 d.f. The difference, $799.07 - 595.84 = 203.23$, will be a measure of the contribution of X_2 to the regression sum of squares, given that X_1 is already present in the regression equation. Then the null hypothesis $\beta_2 = 0$ can be tested by

$$F = \frac{203.23}{24.53} = 8.28$$

with 1 and 17 d.f. From the table of F in the Appendix we find that this is a significant value with $\alpha = .05$.

Similarly, to determine whether X_1 contributes significantly to the regression sum of squares, given that X_2 has already been included in the regression equation, we find the regression sum of squares when only X_2 is included in the regression equation. Using the appropriate values given in Table 14.1, we have

$$SS_{reg} = \frac{(\Sigma x_2 y)^2}{\Sigma x_2^2} = \frac{(364.8)^2}{304.0} = 437.76$$

with 1 d.f. Then the difference, 799.07 − 437.76 = 361.31, will measure the contribution of X_1 to the regression sum of squares, given that X_2 has already been included in the regression equation. In this instance, we have

$$F = \frac{361.31}{24.53} = 14.73$$

with 1 and 17 d.f. and this is also a significant value with $\alpha = .05$. These tests of significance are summarized in Table 14.2.

14.9 Semipartial Correlations and Multiple Correlation

In the preceding section we found that the sum of squares for regression when both X_1 and X_2 were included in the regression equation was equal to 799.07. Using X_1 alone the regression sum of squares was 595.84 and the difference, 799.07 − 595.84 = 203.23, was said to represent the unique contribution of X_2, given that X_1 was already included in the regression equation. If we divide 203.23 by $SS_{tot} = 1216.00$ and take the square root, we have

$$r_{Y(2.1)} = \sqrt{\frac{203.23}{1216.00}} = .4088$$

where $r_{Y(2.1)}$ is called a *semipartial* correlation coefficient. It is, in fact, the correlation coefficient between Y and X_2 after the variance that X_1 has in common with X_2 has been removed from X_2.

The semipartial correlation $r_{Y(2.1)}$ in our three-variable problem will also be given by

$$r_{Y(2.1)} = \frac{r_{2Y} - r_{1Y}r_{12}}{\sqrt{1 - r_{12}^2}} \tag{14.20}$$

TABLE 14.2 Conditional tests of significance for X_1 and X_2

Source	d.f.	Sum of Squares	Mean Square	F
X_1 and X_2	2	799.07		
X_1 alone	1	595.84		
X_2 after X_1	1	203.23	203.23	8.28
X_1 and X_2	2	799.07		
X_2 alone	1	437.76		
X_1 after X_2	1	361.31	361.31	14.73
Residual	17	416.93	24.53	

Substituting the appropriate values of the correlation coefficients in (14.20), we have

$$r_{Y(2.1)} = \frac{.60 - (.70)(.30)}{\sqrt{1 - (.30)^2}} = .4088$$

as before.

To gain a better insight into the nature of the semipartial correlation $r_{Y(2.1)}$ we derive (14.20). Let us assume, without loss of generality, that X_1, X_2, and Y are all in standardized form. Then the regression equation for predicting z_2 from z_1 will be $z_2' = r_{12}z_1$, and we know that the residuals

$$z_2 - z_2' = z_2 - r_{12}z_1$$

will be uncorrelated with z_1 because the covariance

$$c_{(z_2 - z_2')z_1} = \frac{1}{n-1}\Sigma(z_2 - r_{12}z_1)z_1 = r_{12} - r_{12}$$

is equal to zero. The residuals, in other words, will have no variance in common with z_1.

Now let us find the correlation coefficient between the residuals and z_Y. For the numerator of the correlation coefficient we have the covariance

$$c_{(z_2 - z_2')z_Y} = \frac{1}{n-1}\Sigma(z_2 - r_{12}z_1)z_Y = r_{2Y} - r_{1Y}r_{12}$$

We know that $s_{z_Y} = 1.0$ and

$$s_{z_2 - z_2'} = \sqrt{\frac{1}{n-1}\Sigma(z_2 - r_{12}z_1)^2} = \sqrt{1 - r_{12}^2}$$

Then

$$\begin{aligned} r_{(z_2 - z_2')z_Y} &= \frac{c_{(z_2 - z_2')z_Y}}{s_{z_Y}s_{z_2 - z_2'}} \\ &= \frac{r_{2Y} - r_{1Y}r_{12}}{\sqrt{1 - r_{12}^2}} \\ &= r_{Y(2.1)} \end{aligned}$$

The semipartial correlation $r_{Y(2.1)}$, as we have said earlier, is simply the correlation between X_2 and Y after the variance that X_1 has in common with X_2 has been removed from X_2.[4]

We know that $r_{1Y} = .70$ and that $(.70)^2 = .4900$ is the proportion of $\Sigma(Y - \overline{Y})^2$ that can be accounted for by the regression of Y on X_1.

---

[4]Note that the partial correlation coefficient

$$r_{Y2.1} = \frac{r_{2Y} - r_{1Y}r_{12}}{\sqrt{1 - r_{1Y}^2}\sqrt{1 - r_{12}^2}}$$

removes the variance of X_1 from both X_2 and Y. The semipartial correlation coefficient, on the other hand, removes the variance of X_1 only from X_2 and not from Y.

Given that we have taken X_1 into account, the additional proportion of $\Sigma(Y - \overline{Y})^2$ that can be accounted for by X_2 will be given by $r^2_{Y(2.1)} = (.4088)^2 = .1671$. We now note that[5]

$$R^2_{YY'} = R^2_{Y.12} = r^2_{1Y} + r^2_{Y(2.1)}$$

or, in our example,

$$R^2_{Y.12} = .4900 + .1671 = .6571$$

In the same manner, we could show that the semipartial correlation $r_{Y(1.2)}$ will be given by

$$r_{Y(1.2)} = \frac{r_{1Y} - r_{2Y}r_{12}}{\sqrt{1 - r^2_{12}}} \tag{14.21}$$

or, in our example,

$$r_{Y(1.2)} = \frac{.70 - (.60)(.30)}{\sqrt{1 - (.30)^2}} = .5451$$

We found that the regression sum of squares, taking into account X_2 alone, was equal to 437.76. When both X_1 and X_2 were included in the regression equation, the regression sum of squares was equal to 799.07. Then the difference, $799.07 - 437.76 = 361.31$, was said to represent the unique contribution of X_1, given that X_2 had already been included in the regression equation. Thus we also have

$$r_{Y(1.2)} = \sqrt{\frac{361.31}{1216.00}} = .5451$$

Then it will also be true that

$$R^2_{YY'} = R^2_{Y.21} = r^2_{2Y} + r^2_{Y(1.2)}$$

We have r_{2Y} equal to .60 and $r_{Y(1.2)}$ equal to .5451. Consequently,

$$\begin{aligned} R^2_{Y.21} &= (.6000)^2 + (.5451)^2 \\ &= .3600 + .2971 \\ &= .6571 \end{aligned}$$

[5]With three X variables, it will also be true that

$$R^2_{Y.123} = r^2_{1Y} + r^2_{Y(2.1)} + r^2_{Y(3.12)}$$

where

$$r_{Y(3.12)} = \frac{r_{Y(3.1)} - r_{Y(2.1)}r_{3(2.1)}}{\sqrt{1 - r^2_{3(2.1)}}}$$

The semipartial correlation $r_{Y(3.12)}$ is the correlation coefficient between Y and X_3 after the variance that X_1 and X_2 have in common with X_3 has been removed from X_3. The same principles are applicable to any number of X variables, but the calculations of the necessary semipartial correlation coefficients become tedious as the number of X variables increases.

When the X variables are correlated, as X_1 and X_2 are in our example, there is simply no satisfactory method of determining the relative contributions of the X variables to the regression sum of squares or the proportion of $\Sigma(Y - \overline{Y})^2$ accounted for by each of the X variables. This will depend, as we have just seen, on the order in which the X variables are entered in the regression equation. When X_1 is the first variable entered, it accounts for $(.70)^2$ or 49 percent of $\Sigma(Y - \overline{Y})^2$. But if X_1 is entered after X_2, then X_1 accounts for only $(.5451)^2$ or 29.71 percent of $\Sigma(Y - \overline{Y})^2$. Similarly, if X_2 is the first variable entered, it accounts for $(.60)^2$ or 36 percent of $\Sigma(Y - \overline{Y})^2$. But if X_2 is entered after X_1, then X_2 will account for only $(.4088)^2$ or 16.71 percent of $\Sigma(Y - \overline{Y})^2$.

When the X variables are correlated, each variable usually accounts for a larger proportion of $\Sigma(Y - \overline{Y})^2$ when it is entered first in the regression equation than when it follows some other variable or variables. The situation is much different when the X variables are mutually orthogonal or uncorrelated, as we shall see in the next section.

14.10 Multiple Correlation When the X Variables Are Mutually Orthogonal

It is important to note, from (14.15), that if the two X variables are uncorrelated, then

$$R^2_{YY'} = r^2_{1Y} + r^2_{2Y}$$

or simply the sum of the squared correlations of X_1 and X_2 with Y. Similar considerations apply to any number of mutually orthogonal or uncorrelated X variables. If k is the number of X variables and if they are mutually orthogonal or uncorrelated, then

$$R^2_{YY'} = r^2_{1Y} + r^2_{2Y} + \cdots + r^2_{kY} \tag{14.22}$$

In this case the contribution of each X variable to the value of $R^2_{YY'}$ is unique and independent of the contribution of every other variable. No matter in which order or in what combination the X variables are entered into the regression equation, the values of r^2_{iY} will give the proportion of the total sum of squares, $\Sigma(Y - \overline{Y})^2$, for the Y variable that can be accounted for by X_i.[6]

[6]It is also true, with uncorrelated X variables, that the regression coefficient

$$b_i = \frac{\Sigma x_i y}{\Sigma x_i^2}$$

and the regression sum of squares

$$SS_{reg} = \frac{(\Sigma x_i y)^2}{\Sigma x_i^2}$$

for any X variable will remain exactly the same no matter in which order or in what combination the X variables are entered into the regression equation.

It is also easy to see, from (14.20) and (14.21), that when there are two X variables, if $r_{12} = 0$, then the semipartial correlations, $r_{Y(2.1)}$ and $r_{Y(1.2)}$, will be equal to r_{2Y} and r_{1Y}, respectively.

14.11 The Regression Coefficients When X_1, X_2, and Y Are in Standardized Form

If X_1, X_2, and Y are in standardized form, then the variance and standard deviation of each of these standardized variables will be equal to one. We note that the regression coefficients as defined by (14.16) and (14.17) will thus be equal to

$$\hat{b}_1 = \frac{r_{1Y} - r_{2Y}r_{12}}{1 - r_{12}^2} \tag{14.23}$$

and

$$\hat{b}_2 = \frac{r_{2Y} - r_{1Y}r_{12}}{1 - r_{12}^2} \tag{14.24}$$

respectively. We use $\hat{b}$ to indicate the regression coefficients for variables in standardized form.

We can, of course, obtain b_1 and b_2 from the values of $\hat{b}_1$ and $\hat{b}_2$ because

$$b_1 = \frac{s_Y}{s_1}\hat{b}_1 \qquad \text{and} \qquad b_2 = \frac{s_Y}{s_2}\hat{b}_2$$

where s_1 and s_2 are the standard deviations of X_1 and X_2, respectively.

14.12 Matrix Calculation of the Regression Coefficients

If k is the number of X variables, then the intercorrelations of these variables can be represented by a $k \times k$ symmetric matrix that we designate as $\mathbf{R}_{ij}$, or the matrix of intercorrelations of the X variables. The k correlations of the X variables with the Y variable can be represented by a $k \times 1$ column vector, which we designate by $\mathbf{R}_{iY}$. Similarly, the k values of the regression coefficients $\hat{b}$ can be represented by a $k \times 1$ column vector, which we designate by $\mathbf{B}_i$.

If we find the inverse of the $\mathbf{R}_{ij}$ matrix, which we designate by $\mathbf{R}_{ij}^{-1}$, then it can be shown that the matrix product.

$$\mathbf{R}_{ij}^{-1}\mathbf{R}_{iY} = \mathbf{B}_i$$

We illustrate the matrix calculations for two independent variables symbolically as follows:

$$\begin{bmatrix} a_{11} & a_{12} \\ a_{21} & a_{22} \end{bmatrix} \begin{bmatrix} r_{1Y} \\ r_{2Y} \end{bmatrix} = \begin{bmatrix} \hat{b}_1 \\ \hat{b}_2 \end{bmatrix}$$

$$\mathbf{R}_{ij}^{-1} \quad \mathbf{R}_{iY} = \mathbf{B}_i$$

where

$$a_{11}r_{1Y} + a_{12}r_{2Y} = \hat{b}_1$$
$$a_{21}r_{1Y} + a_{22}r_{2Y} = \hat{b}_2$$

The inverse of $\mathbf{R}_{ij}$ is defined if $\mathbf{R}_{ij}^{-1}\mathbf{R}_{ij} = \mathbf{I}$, where $\mathbf{I}$ is an identity matrix with 1's in the upper-left to lower-right diagonal and 0's elsewhere or, in the case of a 2×2 matrix, if

$$\begin{bmatrix} a_{11} & a_{12} \\ a_{21} & a_{22} \end{bmatrix} \begin{bmatrix} r_{11} & r_{12} \\ r_{21} & r_{22} \end{bmatrix} = \begin{bmatrix} 1 & 0 \\ 0 & 1 \end{bmatrix}$$

$$\mathbf{R}_{ij}^{-1} \quad \mathbf{R}_{ij} = \mathbf{I}$$

where

$$a_{11}r_{11} + a_{12}r_{21} = 1$$
$$a_{11}r_{12} + a_{12}r_{22} = 0$$
$$a_{21}r_{11} + a_{22}r_{21} = 0$$
$$a_{21}r_{12} + a_{22}r_{22} = 1$$

We note that for the symmetric 2×2 correlation matrix, $r_{11} = r_{22} = 1$ and $r_{21} = r_{12}$, and, in this case, the four equations become

$$a_{11} + a_{12}r_{12} = 1$$
$$a_{11}r_{12} + a_{12} = 0$$
$$a_{21} + a_{22}r_{12} = 0$$
$$a_{21}r_{12} + a_{22} = 1$$

The first two equations include two unknowns, a_{11} and a_{12}, and these can be solved for, in the manner described previously, using standard algebraic methods. Multiplying the second equation by r_{12}, for example, we have

$$a_{11}r_{12}^2 + a_{12}r_{12} = 0$$

and subtracting this equation from the first equation, we have

$$a_{11} - a_{11}r_{12}^2 = 1$$

and

$$a_{11} = \frac{1}{1 - r_{12}^2}$$

Multiplying the first equation by r_{12}, we have

$$a_{11}r_{12} + a_{12}r_{12}^2 = r_{12}$$

and subtracting this equation from the second equation, we have

$$a_{12} - a_{12}r_{12}^2 = -r_{12}$$

and

$$a_{12} = \frac{-r_{12}}{1 - r_{12}^2}$$

Similarly, the second two equations include the two unknowns a_{21} and a_{22}, and following the same procedures we used with the first two equations, we find that

$$a_{21} = \frac{-r_{12}}{1 - r_{12}^2}$$

and

$$a_{22} = \frac{1}{1 - r_{12}^2}$$

In the example considered in this chapter we have two X variables. The 2×2 correlation matrix is as follows:

$$\mathbf{R}_{ij} = \begin{bmatrix} r_{11} & r_{12} \\ r_{21} & r_{22} \end{bmatrix} = \begin{bmatrix} 1.00 & .30 \\ .30 & 1.00 \end{bmatrix}$$

The inverse of this 2×2 matrix will then be given by

$$\mathbf{R}_{ij}^{-1} = \begin{bmatrix} \dfrac{1}{1 - r_{12}^2} & \dfrac{-r_{12}}{1 - r_{12}^2} \\ \dfrac{-r_{21}}{1 - r_{12}^2} & \dfrac{1}{1 - r_{12}^2} \end{bmatrix} = \begin{bmatrix} \dfrac{1}{.91} & \dfrac{-.30}{.91} \\ \dfrac{-.30}{.91} & \dfrac{1}{.91} \end{bmatrix}$$

or

$$\mathbf{R}_{ij}^{-1} = \begin{bmatrix} 1.0989 & -.3297 \\ -.3297 & 1.0989 \end{bmatrix}$$

If $\mathbf{R}_{ij}^{-1}$, as just given, is actually the inverse of $\mathbf{R}_{ij}$, then it will also be true that

$$\mathbf{R}_{ij}^{-1}\mathbf{R}_{ij} = \mathbf{I} = \begin{bmatrix} 1 & 0 \\ 0 & 1 \end{bmatrix}$$

In our example, we have

$$\underset{\mathbf{R}_{ij}^{-1}}{\begin{bmatrix} 1.0989 & -.3297 \\ -.3297 & 1.0989 \end{bmatrix}} \underset{\mathbf{R}_{ij}}{\begin{bmatrix} 1.00 & .30 \\ .30 & 1.00 \end{bmatrix}} \underset{=}{=} \underset{\mathbf{I}}{\begin{bmatrix} 1 & 0 \\ 0 & 1 \end{bmatrix}}$$

Then

$$\begin{bmatrix} 1.0989 & -.3297 \\ -.3297 & 1.0989 \end{bmatrix} \begin{bmatrix} .70 \\ .60 \end{bmatrix} = \begin{bmatrix} .5714 \\ .4286 \end{bmatrix}$$

$$\mathbf{R}_{\mathbf{ij}}^{-1} \qquad \mathbf{R}_{\mathbf{iY}} \qquad \mathbf{B}_{\mathbf{i}}$$

and $\hat{b}_1 = .5714$ and $\hat{b}_2 = .4286$ are the regression coefficients for standardized variables. We have $s_Y = 8$, $s_1 = 2$, and $s_2 = 4$. Thus, for the values of b_1 and b_2, we have

$$b_1 = \frac{s_Y}{s_1}\hat{b}_1 = \frac{8}{2}(.5714) = 2.2856$$

and

$$b_2 = \frac{s_Y}{s_2}\hat{b}_2 = \frac{8}{4}(.4286) = .8572$$

and these values are equal, within rounding errors, to the values of b_1 and b_2 obtained previously.

14.13 Package Programs for Multiple Regression

Almost all, if not all, computer centers have available a standard package program that will do all of the necessary calculations for a multiple regression problem. With a high-speed electronic computer these calculations can be performed in less than a minute or at most a few minutes, even when the number of X variables is large. Although the package programs may vary from one computing center to another, almost all of them will calculate and print out the means and standard deviations of all of the variables, the complete correlation matrix, the values of $R_{YY'}$ and $R^2_{YY'}$, and a test of significance of $R^2_{YY'}$ or, equivalently, a test of significance of the mean square for linear regression. The package programs also provide a number of options with respect to how the data are to be entered into the computer and how the statistics are to be calculated. In this section we illustrate a portion of the output from one package program that is in use at a large number of computer centers.[7]

Table 14.3 gives the values of five X variables and one Y variable for $n = 15$ cases or subjects. We have included two binary or dichotomous X variables, X_2 and X_3, so that r_{23} is a phi coefficient. Note also that the intercorrelations of the X variables involve a number of point biserial coefficients; for example, r_{12}, r_{13}, r_{24}, r_{25}, r_{34}, and r_{35}. We assume that

[7]See Norman H. Nie, Dale H. Bent, and C. Hadlai Hull, *Statistical Package Program for the Social Sciences.* New York: McGraw-Hill, 1970.

TABLE 14.3 Values of X_1, X_2, X_3, X_4, X_5, and Y for $n = 15$ subjects

X_1	X_2	X_3	X_4	X_5	Y
1	1	0	1	0	5
6	1	0	6	0	12
3	1	0	3	0	9
4	1	0	4	0	8
5	1	0	5	0	11
2	0	1	0	2	1
3	0	1	0	3	2
6	0	1	0	6	7
4	0	1	0	4	3
7	0	1	0	7	8
1	0	0	0	0	10
4	0	0	0	0	13
5	0	0	0	0	16
3	0	0	0	0	12
6	0	0	0	0	17

TABLE 14.4 Means and standard deviations of the variables in Table 14.3

Variable	Mean	Standard Deviation
X_1	4.0000	1.8516
X_2	.3333	.4880
X_3	.3333	.4880
X_4	1.2667	2.1202
X_5	1.4667	2.4162
Y	8.9333	4.8028

all of the X variables are fixed. The correlations of the X variables with the Y variable are either the usual standard correlation coefficients or the special case of point biserial coefficients.

Table 14.4 gives the means and standard deviations of the variables as calculated by the package program, and Table 14.5 gives the intercorrelations of all of the variables. The printed output gives $R_{YY'}$ as being equal to .99169 and $R^2_{YY'}$ as being equal to .98344, as shown in the last line of Table 14.7.

Table 14.6 shows the test of significance of the mean square for linear regression when all five X variables are included in the multiple regression equation or, equivalently, of $R^2_{Y.12345}$. The regression sum of squares is equal to 317.58701 with 5 d.f. and the residual sum of squares is equal to 5.34632 with 9 d.f. For the test of significance, we have

$$F = \frac{MS_{reg}}{MS_{res}} = \frac{63.51740}{.59404} = 106.92$$

TABLE 14.5 Intercorrelations of the variables in Table 14.3

Variable	Intercorrelations				
X_2	−.07906				
X_3	.15811	−.50000			
X_4	.20014	.87455	−.43727		
X_5	.41510	−.44429	.88857	−.38855	
Y	.42570	.01016	−.72134	.14918	−.48339
	X_1	X_2	X_3	X_4	X_5

TABLE 14.6 Test of significance of the mean square for linear regression

Source	Sum of Squares	d.f.	Mean Square	F
Regression	317.58701	5	63.51740	106.92519
Residual	5.34632	9	.59404	

Table 14.7 shows the values of $R_{YY'}$ and $R^2_{YY'}$ when different sets of the X variables are included in the regression equation. For example, when only X_1 and X_2 are included in the regression equation, we have $R^2_{Y.12} = .18315$. When X_1, X_2, and X_3 are included in the regression equation, we have $R^2_{Y.123} = .98294$, and this value is almost equal to $R^2_{Y.12345} = .98344$. Obviously, including X_4 and X_5 in the regression equation does not increase $R^2_{YY'}$ by any appreciable amount over the value we obtain when only X_1, X_2, and X_3 are included in the regression equation.[8] This finding suggests that

$$Y' = a + b_1X_1 + b_2X_2 + b_3X_3$$

will adequately predict the Y values.

The problem was, therefore, rerun using the same package program, but this time only the first three X variables were included in the regression equation. In addition, the computer was instructed to calculate and print out the values of the regression coefficients, b_1, b_2, and b_3, and

[8] A test of significance of the difference between $R^2_{Y.12345}$ and $R^2_{Y.123}$ is provided by

$$F = \frac{(R^2_{Y.12345} - R^2_{Y.123})/2}{(1 - R^2_{Y.12345})/(15 - 5 - 1)} = \frac{(.98344 - .98294)/2}{(1 - .98344)/9} = .136$$

a nonsignificant value with 2 and 9 d.f. Note that $R^2_{Y.12345}$ has 5 d.f. and $R^2_{Y.123}$ has 3 d.f. Thus the difference between these two values has $5 - 3 = 2$ d.f.

TABLE 14.7 Values of $R_{YY'}$ and $R^2_{YY'}$ for different sets of X variables

Variables	$R_{YY'}$	$R^2_{YY'}$
X_1 alone	.42570	.18122
X_1 and X_2	.42796	.18315
X_1, X_2, and X_3	.99144	.98294
X_1, X_2, X_3, and X_4	.99168	.98342
X_1, X_2, X_3, X_4, and X_5	.99169	.98344

TABLE 14.8 Test of significance of $R^2_{Y.123}$

Source	Sum of Squares	d.f.	Mean Square	F
Regression	317.42564	3	105.80855	211.32154
Residual	5.50769	11	.50070	

the value of a. Rounding the calculated values to four decimal places, we have $b_1 = 1.4359$, $b_2 = -4.600$, $b_3 = -10.2615$, and $a = 8.1436$. The value of $R^2_{Y.123}$ is equal to .98294, as before.

The test of significance of the mean square for linear regression or, equivalently, of $R^2_{Y.123}$, is shown in Table 14.8. In this instance, we have

$$F = \frac{MS_{reg}}{MS_{res}} = \frac{105.80855}{.50070} = 211.32$$

with 3 and 11 d.f.

It may be of interest to note that the running time on the computer for the calculations involved in both problems, one with five X variables and one with three X variables, was less than one second.

Exercises

14.1. If $r_{1Y} = .70$ and $r_{2Y} = .60$, is it possible for r_{12} to be equal to $-.30$? Explain why or why not.

14.2. If $r_{1Y} = .20$, $r_{2Y} = .40$, and $r_{3Y} = .60$, and if $r_{12} = r_{13} = r_{23} = 0$, then what is the value of $R^2_{YY'}$?

14.3. Assume that we have $n = 20$ observations with

$$\begin{array}{lll} \overline{X}_1 = 4.0 & \Sigma x_1^2 = 76.0 & \Sigma x_1 y = 212.8 \\ \overline{X}_2 = 8.0 & \Sigma x_2^2 = 304.0 & \Sigma x_2 y = 364.8 \\ \overline{Y} = 20.0 & \Sigma y^2 = 1216.0 & \Sigma x_1 x_2 = 0 \end{array}$$

(a) Find the values of b_1 and b_2 in the multiple regression equation

$$Y' = a + b_1X_1 + b_2X_2$$

(b) Find the value of b_1 in the regression equation $Y' = a + b_1X_1$. Is the value of b_1 equal to the value of b_1 obtained in the multiple regression equation? (c) Find the value of b_2 in the regression equation $Y' = a + b_2X_2$. Is this value of b_2 equal to the value of b_2 obtained in the multiple regression equation? (d) Find the value of $R^2_{YY'}$. (e) Is $R^2_{YY'}$ significant with $\alpha = .05$? (f) Find the inverse of the 2×2 correlation matrix of the two independent variables. (g) What proportion of the total sum of squares, SS_{tot}, is accounted for by X_1? (h) What proportion of the total sum of squares is accounted for by X_2? (i) What proportion of the total sum of squares is accounted for by X_1 and X_2?

14.4. Assume that we have $n = 20$ observations with

$$\bar{X}_1 = 4.0 \qquad \Sigma x_1^2 = 76.0 \qquad \Sigma x_1 y = 152.0$$
$$\bar{X}_2 = 8.0 \qquad \Sigma x_2^2 = 304.0 \qquad \Sigma x_2 y = 304.0$$
$$\bar{Y} = 20.0 \qquad \Sigma y^2 = 1216.0 \qquad \Sigma x_1 x_2 = 76.0$$

(a) Find the values of r_{1Y} and r_{2Y}. (b) Are these values of r significant with $\alpha = .05$? (c) Can we conclude that b_1 in the regression equation $Y' = a + b_1X_1$ is significantly different from zero? Explain why or why not. (d) Can we conclude that b_2 in the regression equation $Y' = a + b_2X_2$ is significantly different from zero? Explain why or why not. (e) Find the value of $R^2_{YY'}$ in the multiple regression equation $Y' = a + b_1X_1 + b_2X_2$. (f) Is $R^2_{YY'}$ significant with $\alpha = .05$? (g) Given that X_1 has been included in the regression equation, does X_2 contribute significantly to the reduction in the residual sum of squares? (h) Given that X_2 has been included in the regression equation, does X_1 contribute significantly to the reduction in the residual sum of squares? (i) Find the inverse of the 2×2 correlation matrix of the independent variables. (j) Use the matrix obtained in (i) to find the values of $\hat{b}_1$ and $\hat{b}_2$. (k) What are the values of b_1 and b_2?

14.5. If the largest absolute correlation between Y and any one of k variables is .50, can $R_{YY'}$ be smaller than .50? Explain why or why not.

14.6. What is the difference between the semipartial correlation coefficient $r_{Y(2.1)}$ and the partial correlation coefficient $r_{Y2.1}$?

14.7. Explain why the inverse of an identity matrix is an identity matrix.

14.8. For the data of Table 14.3, we have $R^2_{Y.123} = .98294$. (a) Using the correlation coefficients given in Table 14.5, calculate $R^2_{Y.13}$. (b) Is $R^2_{Y.123}$ significantly larger than $R^2_{Y.13}$? In other words, does X_2 contribute significantly to $R^2_{Y.123}$?

Answers to the Exercises

Chapter 1

1.1. $a = 1.8$ and $b = .4$
1.2. $a = -5.0$ and $b = -.4$
1.3. $a = 10$ and $b = .3$
1.4. $a = 5$ and $b = -.2$
1.5. $a = 10$ and $b = -.5$

Chapter 2

2.1. $Y = 2X^2$

2.2. $Y = 2\frac{1}{\sqrt{X}}$

2.3. $Y = 2\sqrt{X}$

2.4. $Y = 2\frac{1}{X^2}$

2.5. If $Y = a10^{bX}$, then $\log Y = \log a + bX$, and the plot of Y against X on semilogarithmic paper should be linear.
2.6. If $Y = aX^b$, then $\log Y = \log a + b \log X$, and the plot of the Y against X on logarithmic paper should be linear.
2.7. If $Y = a + b \log X$, then the plot of Y against X on semilogarithmic paper should be linear.

Chapter 3

3.1. $a = 3.62$ and $b = .483$
3.2. (a) $Y' = -.82 + 1.11X$
(b) $s^2_{Y.X} = .855$ and $s_{Y.X} = .925$

(c) $\overline{Y} = .40$ and $\overline{X} = 1.10$
(d) $s_Y^2 = 15.60$ and $s_X^2 = 12.10$

Chapter 4

4.1. (a) $\overline{X} = 14.6$ and $\overline{Y} = 9.9$
(b) $s_X^2 = 81.82$ and $s_Y^2 = 36.54$
(c) $r = .89$
(d) $Y' = 1.29 + .59X$
(e) $X' = 1.53 + 1.32Y$
(f) $s_{Y.X}^2 = 8.90$ and $s_{X.Y}^2 = 19.94$

4.2. (a) $r = .89$
(b) $b_Y = 1.00$ and $b_X = .79$

4.4. We have $Y - Y' = y - bx$. Then the numerator of the correlation coefficient will be

$$\Sigma(y - bx)x = \Sigma xy - b\Sigma x^2$$

Substituting $\Sigma xy/\Sigma x^2$ for b in the preceding expression, we see that the numerator of the correlation coefficient is equal to zero.

4.6. $b_X = .75$ and $b_Y = .48$

4.7. $\Sigma(Y - Y')^2 = 0$

4.8. No, because $b_Y b_X = r^2$ and r^2 cannot be greater than 1.00.

4.9. If $r = 1.00$, then $b_X = .50$

4.10. No, because b_X and b_Y must have the same sign.

Chapter 5

5.1. If $X = 43$, then $z = 2.0$

5.7. (a) If $z = 0$, then IQ = 100
(b) If $z = 1.0$, then IQ = 115
(c) If $z = 2.0$, then IQ = 130
(d) If $z = -1.0$, then IQ = 85
(e) If $z = -2.0$, then IQ = 70

5.8. For the correlation coefficient between z_1' and z_1, we have

$$r_{z_1' z_1} = \frac{c_{z_1' z_1}}{s_{z_1'} s_{z_1}} \qquad (1)$$

For the proof $r_{z_1' z_1} = r$, we note that

$$c_{z_1' z_1} = \frac{\Sigma z_1' z_1}{n - 1} = \frac{r \Sigma z_1 z_2}{n - 1} = r^2$$

We proved in the text that $s_{z_1'}^2 = r^2$ and, consequently, $s_{z_1'} = r$. We also proved that for a set of standard scores $s_z = 1$. Then, substituting in (1), we have

$$r_{z_1' z_1} = \frac{r^2}{r} = r$$

Chapter 6

6.5. (a) No, because $r_{12.3} = 2.53$ and this is an impossible value.
(b) $r_{12} = .02$
(c) $r_{12} = -1.00$

6.8. (a) $\frac{\Sigma z'_1}{n} = \frac{\Sigma(r_{12}z_2 + r_{13}z_3)}{n} = 0$

$$s^2_{z'_1} = \frac{\Sigma(r_{12}z_2 + r_{13}z_3)^2}{n-1} = 1.00$$

(b) $r_{z'_1 z_1} = \sqrt{r^2_{12} + r^2_{13}} = 1.00$

Chapter 7

7.1. $r = .23$
7.2. $r = .63$
7.3. $r = -.10$
7.4. $r = .13$
7.5. $r = .30$

Chapter 8

8.1. $r \geq .632$ or $r \leq -.632$ would be significant
8.2. $r \geq .44$ or $r \leq -.44$ would be significant
8.3. $.63 < \rho < .92$
8.4. $Z = .685$ and $\chi^2 = .470$
8.5. (a) $\chi^2 = 7.611$ with 2 d.f. is a significant value with $\alpha = .05$.
(b) The estimate would not be obtained because the evidence indicates that the samples are not from a common population.

Chapter 9

9.1. (a) .18, .42, .12, and .28
(b) 9, 21, 6, and 14
(c) $\chi^2 = 19.44$ with 1 d.f.
9.2. (a) $t = 3.0$ with 14 d.f.
(b) $F = 9.0$ with 1 and 14 d.f.
9.3. Yes, either from Table IX or from Table V.
9.4. $\chi^2 = (200)(.23)^2 = 10.58$ with 1 d.f.

Chapter 10

10.1. (a) $b = 2.0$
(b) $s^2_{Y.X} = 1.3333$ and $s_b = .365$
(c) $t = 5.48$ with 3 d.f.

10.2. (a) $b = .5$
(b) $s^2_{Y.X} = .5000$ and $s_b = .224$
(c) $t = 2.236$ with 3 d.f.

10.3. (a) $s^2_{Y.X} = .9167$
(b) $s_{b_1-b_2} = .4282$
(c) $t = 3.503$ with 6 d.f.

10.4. (a) $SS_1 = 5.50$, $SS_2 = 16.75$, $SS_3 = 11.25$
(b) $F = 12.27$ with 1 and 6 d.f.
(c) $t^2 = (3.503)^2 = 12.27$

10.8. We have

$$t = \frac{b}{\sqrt{s^2_{Y.X}/\Sigma x^2}}$$

Multiplying the numerator and the denominator of the right side of the equation by $\sqrt{\Sigma x^2}$ and substituting identities for b and $s^2_{Y.X}$, we obtain

$$t = \frac{\dfrac{\Sigma xy \sqrt{\Sigma x^2}}{\Sigma x^2}}{\sqrt{\left[\Sigma y^2 - \dfrac{(\Sigma xy)^2}{\Sigma x^2}\right] \Big/ (n-2)}}$$

But $\Sigma x^2 = \sqrt{\Sigma x^2}\sqrt{\Sigma x^2}$. Substituting this identity in the numerator of the preceding expression and multiplying both the numerator and the denominator by $1/\sqrt{\Sigma y^2}$, we obtain

$$t = \frac{r}{\sqrt{1 - r^2}}\sqrt{n-2}$$

Chapter 11

11.1. (a) $SS_T = 50.0$
$SS_L = 50.0$

11.2. (a) $SS_T = 70.0$
$SS_Q = 70.0$

11.3. (a) $SS_T = 50.0$
$SS_C = 50.0$

Chapter 12

12.1. (a) $SS_T = 106.00$, $SS_W = 50.00$, $F = 10.60$ with 4 and 20 d.f.
(b) $b_1 = 1.100$, $b_2 = -.357$, $b_3 = .800$, $b_4 = .114$
(c) Linear $= 60.50$, $F = 24.20$ with 1 and 20 d.f.
Quadratic $= 8.93$, $F = 3.57$ with 1 and 20 d.f.
Cubic $= 32.00$, $F = 12.80$ with 1 and 20 d.f.
Quartic $= 4.57$, $F = 1.83$ with 1 and 20 d.f.
(d) $SS_{res} = 106.00 - 60.50 = 45.50$
$F = 6.07$ with 3 and 20 d.f.

(e) $SS_{res} = 106.00 - 60.50 - 8.93 = 36.57$
$F = 7.31$ with 2 and 20 d.f.
(f) $SS_{res} = 106.00 - 60.50 - 8.93 - 32.00 = 4.57$
$F = 1.83$ with 1 and 20 d.f.
(g) .571, .084, .302, .043
(h) $.571 + .302 = .873$

12.2. If $\overline{Y}_i' = \overline{Y} + b_1 x_1$, then

$$n\Sigma(\overline{Y}_i - \overline{Y}_i')^2 = n\Sigma[(\overline{Y}_i - \overline{Y}) - b_1 x_1]^2$$
$$= n[\Sigma(\overline{Y}_i - \overline{Y})^2 - 2b_1\Sigma x_1(\overline{Y}_i - \overline{Y}) + b_1^2\Sigma x_1^2]$$

We have shown that

$$\Sigma x_1(\overline{Y}_i - \overline{Y}) = \frac{1}{n}\Sigma(x_1\Sigma Y_i) \text{ and } b_1 = \frac{\Sigma(x_1\Sigma Y_i)}{n\Sigma x_1^2}$$

Then

$$n\Sigma(\overline{Y}_i - \overline{Y}_i')^2 = n\Sigma(\overline{Y}_i - \overline{Y})^2 - 2\,\frac{[\Sigma(x_1\Sigma Y_i)]^2}{n\Sigma x_1^2} + \frac{[\Sigma(x_1\Sigma Y_i)]^2}{n\Sigma x_1^2}$$

The first term on the right is the treatment sum of squares, and we also have

$$SS_L = \frac{[\Sigma(x_1\Sigma Y_i)]^2}{n\Sigma x_1^2}$$

so that

$$n\Sigma(\overline{Y}_i - \overline{Y}_i')^2 = SS_T - SS_L$$

Chapter 13

13.1. (a) $MS_W = 2.5$, $MS_T = 5.0$, and $F = MS_T/MS_W = 2.0$
(b) $SS_L = 2.5$ and $SS_Q = 7.5$
(c) Yes
(d) $F = 2.5/2.5 = 1.0$ with 1 and 12 d.f.
(e) $F = 7.5/2.5 = 3.0$ with 1 and 12 d.f.

13.2. (a) $SS_S = 20.0$ and $SS_{ST} = 10.0$
(b) $F = MS_T/MS_{ST} = 5.0/1.25 = 4.0$ with 2 and 8 d.f.
(c) $F = 2.5/1.25 = 2.0$ with 1 and 8 d.f.
(d) $F = 7.5/1.25 = 6.0$ with 1 and 8 d.f.
(e) $c_{12} = c_{13} = c_{23} = 1.25$
(f) $\bar{c}_{ij} = 1.25$
(g) $MS_{ST} = MS_W - \bar{c}_{ij} = 2.50 - 1.25 = 1.25$

13.3. (a) $\overline{Y}_1' = 6.68$, $\overline{Y}_2' = 5.56$, $\overline{Y}_3' = 4.44$, $\overline{Y}_4' = 3.32$
(b) $r = .9719$
(c) $r^2 = .9446$

13.4. (a) $\overline{Y}_1' = 6.98$, $\overline{Y}_2' = 5.26$, $\overline{Y}_3' = 4.14$, $\overline{Y}_4' = 3.62$
(b) $r = .9994$
(c) $.9446 + .0542 = .9988$

13.5. (a) $\overline{Y}_1' = 7.0$, $\overline{Y}_2' = 5.2$, $\overline{Y}_3' = 4.2$, $\overline{Y}_4' = 3.6$
(b) $r = 1.0000$
(c) $.9446 + .0542 + .0012 = 1.0000$

13.6. (a) MS_W will have 45 d.f. and MS_{ST} will have 36 d.f.
(b) If the same subjects are tested on each of the 5 days, carry-over effects may influence the values of the means. For example, recall on the second day may be greater in the repeated measure design than in the completely randomized design because the subjects will have experienced one recall 24 hours earlier and this may influence the amount recalled on the second day. Similar considerations apply to the other successive test periods.

Chapter 14

14.1. No. If $r = -.30$, then $R^2_{YY'} = 1.21$ and this is an impossible value.

14.2 $R^2_{YY'} = (.20)^2 + (.40)^2 + (.60)^2 = .56$

14.3. (a) $b_1 = 2.8$ and $b_2 = 1.2$
(b) $b_1 = 2.8$
(c) $b_2 = 1.2$
(d) $R^2_{YY'} = .85$
(e) $F = 48.17$ with 2 and 17 d.f.
(f) $\mathbf{R}_{ij}^{-1} = \begin{bmatrix} 1 & 0 \\ 0 & 1 \end{bmatrix}$
(g) .49
(h) .36
(i) .85

14.4. (a) $r_{1Y} = .50$ and $r_{2Y} = .50$
(b) Yes
(c) Yes
(d) Yes
(e) $R^2_{YY'} = .3333$
(f) $F = 4.25$ with 2 and 17 d.f.
(g) $F = 2.13$ with 1 and 17 d.f.
(h) $F = 2.13$ with 1 and 17 d.f.
(i) $\mathbf{R}_{ij}^{-1} = \begin{bmatrix} 1.3333 & -.6667 \\ -.6667 & 1.3333 \end{bmatrix}$
(j) $\hat{b}_1 = .3333$ and $\hat{b}_2 = .3333$
(k) $b_1 = 1.3332$ and $b_2 = .6666$

14.8. (a) $R^2_{Y.13} = .81913$
(b) $F = 105.68$ with 1 and 11 d.f.

Appendix

TABLE I Table of squares, square roots, and reciprocals of numbers from 1 to 1000

N	N^2	$\sqrt{N}$	$1/N$	N	N^2	$\sqrt{N}$	$1/N$
1	1	1.0000	1.000000	46	2116	6.7823	.021739
2	4	1.4142	.500000	47	2209	6.8557	.021277
3	9	1.7321	.333333	48	2304	6.9282	.020833
4	16	2.0000	.250000	49	2401	7.0000	.020408
5	25	2.2361	.200000	50	2500	7.0711	.020000
6	36	2.4495	.166667	51	2601	7.1414	.019608
7	49	2.6458	.142857	52	2704	7.2111	.019231
8	64	2.8284	.125000	53	2809	7.2801	.018868
9	81	3.0000	.111111	54	2916	7.3485	.018519
10	100	3.1623	.100000	55	3025	7.4162	.018182
11	121	3.3166	.090909	56	3136	7.4833	.017857
12	144	3.4641	.083333	57	3249	7.5498	.017544
13	169	3.6056	.076923	58	3364	7.6158	.017241
14	196	3.7417	.071429	59	3481	7.6811	.016949
15	225	3.8730	.066667	60	3600	7.7460	.016667
16	256	4.0000	.062500	61	3721	7.8102	.016393
17	289	4.1231	.058824	62	3844	7.8740	.016129
18	324	4.2426	.055556	63	3969	7.9373	.015873
19	361	4.3589	.052632	64	4096	8.0000	.015625
20	400	4.4721	.050000	65	4225	8.0623	.015385
21	441	4.5826	.047619	66	4356	8.1240	.015152
22	484	4.6904	.045455	67	4489	8.1854	.014925
23	529	4.7958	.043478	68	4624	8.2462	.014706
24	576	4.8990	.041667	69	4761	8.3066	.014493
25	625	5.0000	.040000	70	4900	8.3666	.014286
26	676	5.0990	.038462	71	5041	8.4261	.014085
27	729	5.1962	.037037	72	5184	8.4853	.013889
28	784	5.2915	.035714	73	5329	8.5440	.013699
29	841	5.3852	.034483	74	5476	8.6023	.013514
30	900	5.4772	.033333	75	5625	8.6603	.013333
31	961	5.5678	.032258	76	5776	8.7178	.013158
32	1024	5.6569	.031250	77	5929	8.7750	.012987
33	1089	5.7446	.030303	78	6084	8.8318	.012821
34	1156	5.8310	.029412	79	6241	8.8882	.012658
35	1225	5.9161	.028571	80	6400	8.9443	.012500
36	1296	6.0000	.027778	81	6561	9.0000	.012346
37	1369	6.0828	.027027	82	6724	9.0554	.012195
38	1444	6.1644	.026316	83	6889	9.1104	.012048
39	1521	6.2450	.025641	84	7056	9.1652	.011905
40	1600	6.3246	.025000	85	7225	9.2195	.011765
41	1681	6.4031	.024390	86	7396	9.2736	.011628
42	1764	6.4807	.023810	87	7569	9.3274	.011494
43	1849	6.5574	.023256	88	7744	9.3808	.011364
44	1936	6.6332	.022727	89	7921	9.4340	.011236
45	2025	6.7082	.022222	90	8100	9.4868	.011111

Portions of Table I have been reproduced from J. W. Dunlap and A. K. Kurtz. *Handbook of Statistical Nomographs, Tables, and Formulas,* World Book Company, New York (1932), by permission of the authors and publishers.

N	N^2	$\sqrt{N}$	$1/N$	N	N^2	$\sqrt{N}$	$1/N$
91	8281	9.5394	.010989	141	19881	11.8743	.00709220
92	8464	9.5917	.010870	142	20164	11.9164	.00704225
93	8649	9.6437	.010753	143	20449	11.9583	.00699301
94	8836	9.6954	.010638	144	20736	12.0000	.00694444
95	9025	9.7468	.010526	145	21025	12.0416	.00689655
96	9216	9.7980	.010417	146	21316	12.0830	.00684932
97	9409	9.8489	.010309	147	21609	12.1244	.00680272
98	9604	9.8995	.010204	148	21904	12.1655	.00675676
99	9801	9.9499	.010101	149	22201	12.2066	.00671141
100	10000	10.0000	.010000	150	22500	12.2474	.00666667
101	10201	10.0499	.00990099	151	22801	12.2882	.00662252
102	10404	10.0995	.00980392	152	23104	12.3288	.00657895
103	10609	10.1489	.00970874	153	23409	12.3693	.00653595
104	10816	10.1980	.00961538	154	23716	12.4097	.00649351
105	11025	10.2470	.00952381	155	24025	12.4499	.00645161
106	11236	10.2956	.00943396	156	24336	12.4900	.00641026
107	11449	10.3441	.00934579	157	24649	12.5300	.00636943
108	11664	10.3923	.00925926	158	24964	12.5698	.00632911
109	11881	10.4403	.00917431	159	25281	12.6095	.00628931
110	12100	10.4881	.00909091	160	25600	12.6491	.00625000
111	12321	10.5357	.00900901	161	25921	12.6886	.00621118
112	12544	10.5830	.00892857	162	26244	12.7279	.00617284
113	12769	10.6301	.00884956	163	26569	12.7671	.00613497
114	12996	10.6771	.00877193	164	26896	12.8062	.00609756
115	13225	10.7238	.00869565	165	27225	12.8452	.00606061
116	13456	10.7703	.00862069	166	27556	12.8841	.00602410
117	13689	10.8167	.00854701	167	27889	12.9228	.00598802
118	13924	10.8628	.00847458	168	28224	12.9615	.00595238
119	14161	10.9087	.00840336	169	28561	13.0000	.00591716
120	14400	10.9545	.00833333	170	28900	13.0384	.00588235
121	14641	11.0000	.00826446	171	29241	13.0767	.00584795
122	14884	11.0454	.00819672	172	29584	13.1149	.00581395
123	15129	11.0905	.00813008	173	29929	13.1529	.00578035
124	15376	11.1355	.00806452	174	30276	13.1909	.00574713
125	15625	11.1803	.00800000	175	30625	13.2288	.00571429
126	15876	11.2250	.00793651	176	30976	13.2665	.00568182
127	16129	11.2694	.00787402	177	31329	13.3041	.00564972
128	16384	11.3137	.00781250	178	31684	13.3417	.00561798
129	16641	11.3578	.00775194	179	32041	13.3791	.00558659
130	16900	11.4018	.00769231	180	32400	13.4164	.00555556
131	17161	11.4455	.00763359	181	32761	13.4536	.00552486
132	17424	11.4891	.00757576	182	33124	13.4907	.00549451
133	17689	11.5326	.00751880	183	33489	13.5277	.00546448
134	17956	11.5758	.00746269	184	33856	13.5647	.00543478
135	18225	11.6190	.00740741	185	34225	13.6015	.00540541
136	18496	11.6619	.00735294	186	34596	13.6382	.00537634
137	18769	11.7047	.00729927	187	34969	13.6748	.00534759
138	19044	11.7473	.00724638	188	35344	13.7113	.00531915
139	19321	11.7893	.00719424	189	35721	13.7477	.00529101
140	19600	11.8322	.00714286	190	36100	13.7840	.00526316

(Continued)

N	N^2	$\sqrt{N}$	$1/N$	N	N^2	$\sqrt{N}$	$1/N$
191	36481	13.8203	.00523560	241	58081	15.5242	.00414938
192	36864	13.8564	.00520833	242	58564	15.5563	.00413223
193	37249	13.8924	.00518135	243	59049	15.5885	.00411523
194	37636	13.9284	.00515464	244	59536	15.6205	.00409836
195	38025	13.9642	.00512821	245	60025	15.6525	.00408163
196	38416	14.0000	.00510204	246	60516	15.6844	.00406504
197	38809	14.0357	.00507614	247	61009	15.7162	.00404858
198	39204	14.0712	.00505051	248	61504	15.7480	.00403226
199	39601	14.1067	.00502513	249	62001	15.7797	.00401606
200	40000	14.1421	.00500000	250	62500	15.8114	.00400000
201	40401	14.1774	.00497512	251	63001	15.8430	.00398406
202	40804	14.2127	.00495050	252	63504	15.8745	.00396825
203	41209	14.2478	.00492611	253	64009	15.9060	.00395257
204	41616	14.2829	.00490196	254	64516	15.9374	.00393701
205	42025	14.3178	.00487805	255	65025	15.9687	.00392157
206	42436	14.3527	.00485437	256	65536	16.0000	.00390625
207	42849	14.3875	.00483092	257	66049	16.0312	.00389105
208	43264	14.4222	.00480769	258	66564	16.0624	.00387597
209	43681	14.4568	.00478469	259	67081	16.0935	.00386100
210	44100	14.4914	.00476190	260	67600	16.1245	.00384615
211	44521	14.5258	.00473934	261	68121	16.1555	.00383142
212	44944	14.5602	.00471698	262	68644	16.1864	.00381679
213	45369	14.5945	.00469484	263	69169	16.2173	.00380228
214	45796	14.6287	.00467290	264	69696	16.2481	.00378788
215	46225	14.6629	.00465116	265	70225	16.2788	.00377358
216	46656	14.6969	.00462963	266	70756	16.3095	.00375940
217	47089	14.7309	.00460829	267	71289	16.3401	.00374532
218	47524	14.7648	.00458716	268	71824	16.3707	.00373134
219	47961	14.7986	.00456621	269	72361	16.4012	.00371747
220	48400	14.8324	.00454545	270	72900	16.4317	.00370370
221	48841	14.8661	.00452489	271	73441	16.4621	.00369004
222	49284	14.8997	.00450450	272	73984	16.4924	.00367647
223	49729	14.9332	.00448430	273	74529	16.5227	.00366300
224	50176	14.9666	.00446429	274	75076	16.5529	.00364964
225	50625	15.0000	.00444444	275	75625	16.5831	.00363636
226	51076	15.0333	.00442478	276	76176	16.6132	.00362319
227	51529	15.0665	.00440529	277	76729	16.6433	.00361011
228	51984	15.0997	.00438596	278	77284	16.6733	.00359712
229	52441	15.1327	.00436681	279	77841	16.7033	.00358423
230	52900	15.1658	.00434783	280	78400	16.7332	.00357143
231	53361	15.1987	.00432900	281	78961	16.7631	.00355872
232	53824	15.2315	.00431034	282	79524	16.7929	.00354610
233	54289	15.2643	.00429185	283	80089	16.8226	.00353357
234	54756	15.2971	.00427350	284	80656	16.8523	.00352113
235	55225	15.3297	.00425532	285	81225	16.8819	.00350877
236	55696	15.3623	.00423729	286	81796	16.9115	.00349650
237	56169	15.3948	.00421941	287	82369	16.9411	.00348432
238	56644	15.4272	.00420168	288	82944	16.9706	.00347222
239	57121	15.4596	.00418410	289	83521	17.0000	.00346021
240	57600	15.4919	.00416667	290	84100	17.0294	.00344828

N	N^2	$\sqrt{N}$	$1/N$	N	N^2	$\sqrt{N}$	$1/N$
291	84681	17.0587	.00343643	341	116281	18.4662	.00293255
292	85264	17.0880	.00342466	342	116964	18.4932	.00292398
293	85849	17.1172	.00341297	343	117649	18.5203	.00291545
294	86436	17.1464	.00340136	344	118336	18.5372	.00290698
295	87025	17.1756	.00338983	345	119025	18.5742	.00289855
296	87616	17.2047	.00337838	346	119716	18.6011	.00289017
297	88209	17.2337	.00336700	347	120409	18.6279	.00288184
298	88804	17.2627	.00335570	348	121104	18.6548	.00287356
299	89401	17.2916	.00334448	349	121801	18.6815	.00286533
300	90000	17.3205	.00333333	350	122500	18.7083	.00285714
301	90601	17.3494	.00332226	351	123201	18.7350	.00284900
302	91204	17.3781	.00331126	352	123904	18.7617	.00284091
303	91809	17.4069	.00330033	353	124609	18.7883	.00283286
304	92416	17.4356	.00328947	354	125316	18.8149	.00282486
305	93025	17.4642	.00327869	355	126025	18.8414	.00281690
306	93636	17.4929	.00326797	356	126736	18.8680	.00280899
307	94249	17.5214	.00325733	357	127449	18.8944	.00280112
308	94864	17.5499	.00324675	358	128164	18.9209	.00279330
309	95481	17.5784	.00323625	359	128881	18.9473	.00278552
310	96100	17.6068	.00322581	360	129000	18.9737	.00277778
311	96721	17.6352	.00321543	361	130321	19.0000	.00277008
312	97344	17.6635	.00320513	362	131044	19.0263	.00276243
313	97969	17.6918	.00319489	363	131769	19.0526	.00275482
314	98596	17.7200	.00318471	364	132496	19.0788	.00274725
315	99225	17.7482	.00317460	365	133225	19.1050	.00273973
316	99856	17.7764	.00316456	366	133956	19.1311	.00273224
317	100489	17.8045	.00315457	367	134689	19.1572	.00272480
318	101124	17.8326	.00314465	368	135424	19.1833	.00271739
319	101761	17.8606	.00313480	369	136161	19.2094	.00271003
320	102400	17.8885	.00312500	370	136900	19.2354	.00270270
321	103041	17.9165	.00311526	371	137641	19.2614	.00269542
322	103684	17.9444	.00310559	372	138384	19.2873	.00268817
323	104329	17.9722	.00309598	373	139129	19.3132	.00268097
324	104976	18.0000	.00308642	374	139876	19.3391	.00267380
325	105625	18.0278	.00307692	375	140625	19.3649	.00266667
326	106276	18.0555	.00306748	376	141376	19.3907	.00265957
327	106929	18.0831	.00305810	377	142129	19.4165	.00265252
328	107584	18.1108	.00304878	378	142884	19.4422	.00264550
329	108241	18.1384	.00303951	379	143641	19.4679	.00263852
330	108900	18.1659	.00303030	380	144400	19.4936	.00263158
331	109561	18.1934	.00302115	381	145161	19.5192	.00262467
332	110224	18.2209	.00301205	382	145924	19.5448	.00261780
333	110889	18.2483	.00300300	383	146689	19.5704	.00261097
334	111556	18.2757	.00299401	384	147456	19.5959	.00260417
335	112225	18.3030	.00298507	385	148225	19.6214	.00259740
336	112896	18.3303	.00297619	386	148996	19.6469	.00259067
337	113569	18.3576	.00296736	387	149769	19.6723	.00258398
338	114244	18.3848	.00295858	388	150544	19.6977	.00257732
339	114921	18.4120	.00294985	389	151321	19.7231	.00257069
340	115600	18.4391	.00294118	390	152100	19.7484	.00256410

(Continued)

N	N^2	$\sqrt{N}$	$1/N$	N	N^2	$\sqrt{N}$	$1/N$
391	152881	19.7737	.00255754	441	194481	21.0000	.00226757
392	153664	19.7990	.00255102	442	195364	21.0238	.00226244
393	154449	19.8242	.00254453	443	196249	21.0476	.00225734
394	155236	19.8494	.00253807	444	197136	21.0713	.00225225
395	156025	19.8746	.00253165	445	198025	21.0950	.00224719
396	156816	19.8997	.00252525	446	198916	21.1187	.00224215
397	157609	19.9249	.00251889	447	199809	21.1424	.00223714
398	158404	19.9499	.00251256	448	200704	21.1660	.00223214
399	159201	19.9750	.00250627	449	201601	21.1896	.00222717
400	160000	20.0000	.00250000	450	202500	21.2132	.00222222
401	160801	20.0250	.00249377	451	203401	21.2368	.00221729
402	161604	20.0499	.00248756	452	204304	21.2603	.00221239
403	162409	20.0749	.00248139	453	205209	21.2838	.00220751
404	163216	20.0998	.00247525	454	206116	21.3073	.00220264
405	164025	20.1246	.00246914	455	207025	21.3307	.00219780
406	164836	20.1494	.00246305	456	207936	21.3542	.00219298
407	165649	20.1742	.00245700	457	208849	21.3776	.00218818
408	166464	20.1990	.00245098	458	209764	21.4009	.00218341
409	167281	20.2237	.00244499	459	210681	21.4243	.00217865
410	168100	20.2485	.00243902	460	211600	21.4476	.00217391
411	168921	20.2731	.00243309	461	212521	21.4709	.00216920
412	169744	20.2978	.00242718	462	213444	21.4942	.00216450
413	170569	20.3224	.00242131	463	214369	21.5174	.00215983
414	171396	20.3470	.00241546	464	215296	21.5407	.00215517
415	172225	20.3715	.00240964	465	216225	21.5639	.00215054
416	173056	20.3961	.00240385	466	217156	21.5870	.00214592
417	173889	20.4206	.00239808	467	218089	21.6102	.00214133
418	174724	20.4450	.00239234	468	219024	21.6333	.00213675
419	175561	20.4695	.00238663	469	219961	21.6564	.00213220
420	176400	20.4939	.00238095	470	220900	21.6795	.00212766
421	177241	20.5183	.00237530	471	221841	21.7025	.00212314
422	178084	20.5426	.00236967	472	222784	21.7256	.00211864
423	178929	20.5670	.00236407	472	223729	21.7486	.00211416
424	179776	20.5913	.00235849	474	224676	21.7715	.00210970
425	180625	20.6155	.00235294	475	225625	21.7945	.00210526
426	181476	20.6398	.00234742	476	226576	21.8174	.00210084
427	182329	20.6640	.00234192	477	227529	21.8403	.00209644
428	183184	20.6882	.00233645	478	228484	21.8632	.00209205
429	184041	20.7123	.00233100	479	229441	21.8861	.00208768
430	184900	20.7364	.00232558	480	230400	21.9089	.00208333
431	185761	20.7605	.00232019	481	231361	21.9317	.00207900
432	186624	20.7846	.00231481	482	232324	21.9545	.00207469
433	187489	20.8087	.00230947	483	233289	21.9773	.00207039
434	188356	20.8327	.00230415	484	234256	22.0000	.00206612
435	189225	20.8567	.00229885	485	235225	22.0227	.00206186
436	190096	20.8806	.00229358	486	236196	22.0454	.00205761
437	190969	20.9045	.00228833	487	237169	22.0681	.00205339
438	191844	20.9284	.00228311	488	238144	22.0907	.00204918
439	192721	20.9523	.00227790	489	239121	22.1133	.00204499
440	193600	20.9762	.00227273	490	240100	22.1359	.00204082

N	N^2	$\sqrt{N}$	$1/N$	N	N^2	$\sqrt{N}$	$1/N$
491	241081	22.1585	.00203666	541	292681	23.2594	.00184843
492	242064	22.1811	.00203252	542	293764	23.2809	.00184502
493	243049	22.2036	.00202840	543	294849	23.3024	.00184162
494	244036	22.2261	.00202429	544	295936	23.3238	.00183824
495	245025	22.2486	.00202020	545	297025	23.3452	.00183486
496	246016	22.2711	.00201613	546	298116	23.3666	.00183150
497	247009	22.2935	.00201207	547	299209	23.3880	.00182815
498	248004	22.3159	.00200803	548	300304	23.4094	.00182482
499	249001	22.3383	.00200401	549	301401	23.4307	.00182149
500	250000	22.3607	.00200000	550	302500	23.4521	.00181818
501	251001	22.3830	.00199601	551	303601	23.4734	.00181488
502	252004	22.4054	.00199203	552	304704	23.4947	.00181159
503	253009	22.4277	.00198807	553	305809	23.5160	.00180832
504	254016	22.4499	.00198413	554	306916	23.5372	.00180505
505	255025	22.4722	.00198020	555	308025	23.5584	.00180180
506	256036	22.4944	.00197628	556	309136	23.5797	.00179856
507	257049	22.5167	.00197239	557	310249	23.6008	.00179533
508	258064	22.5389	.00196850	558	311364	23.6220	.00179211
509	259081	22.5610	.00196464	559	312481	23.6432	.00178891
510	260100	22.5832	.00196078	560	313600	23.6643	.00178571
511	261121	22.6053	.00195695	561	314721	23.6854	.00178253
512	262144	22.6274	.00195312	562	315844	23.7065	.00177936
513	263169	22.6495	.00194932	563	316969	23.7276	.00177620
514	264196	22.6716	.00194553	564	318096	23.7487	.00177305
515	265225	22.6936	.00194175	565	319225	23.7697	.00176991
516	266256	22.7156	.00193798	566	320356	23.7908	.00176678
517	267289	22.7376	.00193424	567	321489	23.8118	.00176367
518	268324	22.7596	.00193050	568	322624	23.8328	.00176056
519	269361	22.7816	.00192678	569	323761	23.8537	.00175747
520	270400	22.8035	.00192308	570	324900	23.8747	.00175439
521	271441	22.8254	.00191939	571	326041	23.8956	.00175131
522	272484	22.8473	.00191571	572	327184	23.9165	.00174825
523	273529	22.8692	.00191205	573	328329	23.9374	.00174520
524	274576	22.8910	.00190840	574	329476	23.9583	.00174216
525	275625	22.9129	.00190476	575	330625	23.9792	.00173913
526	276676	22.9347	.00190114	576	331776	24.0000	.00173611
527	277729	22.9565	.00189753	577	332929	24.0208	.00173310
528	278784	22.9783	.00189394	578	334084	24.0416	.00173010
529	279841	23.0000	.00189036	579	335241	24.0624	.00172712
530	280900	23.0217	.00188679	580	336400	24.0832	.00172414
531	281961	23.0434	.00188324	581	337561	24.1039	.00172117
532	283024	23.0651	.00187970	582	338724	24.1247	.00171821
533	284089	23.0868	.00187617	583	339889	24.1454	.00171527
534	285156	23.1084	.00187266	584	341056	24.1661	.00171233
535	286225	23.1301	.00186916	585	342225	24.1868	.00170940
536	287296	23.1517	.00186567	586	343396	24.2074	.00170648
537	288369	23.1733	.00186220	587	344569	24.2281	.00170358
538	289444	23.1948	.00185874	588	345744	24.2487	.00170068
539	290521	23.2164	.00185529	589	346921	24.2693	.00169779
540	291600	23.2379	.00185185	590	348100	24.2899	.00169492

(Continued)

N	N^2	$\sqrt{N}$	$1/N$	N	N^2	$\sqrt{N}$	$1/N$
591	349281	24.3105	.00169205	641	410881	25.3180	.00156006
592	350464	24.3311	.00168919	642	412164	25.3377	.00155763
593	351649	24.3516	.00168634	643	413449	25.3574	.00155521
594	352836	24.3721	.00168350	644	414736	25.3772	.00155280
595	354025	24.3926	.00168067	645	416025	25.3969	.00155039
596	355216	24.4131	.00167785	646	417316	25.4165	.00154799
597	356409	24.4336	.00167504	647	418609	25.4362	.00154560
598	357604	24.4540	.00167224	648	419904	25.4558	.00154321
599	358801	24.4745	.00166945	649	421201	25.4755	.00154083
600	360000	24.4949	.00166667	650	422500	25.4951	.00153846
601	361201	24.5153	.00166389	651	423801	25.5147	.00153610
602	362404	24.5357	.00166113	652	425104	25.5343	.00153374
603	363609	24.5561	.00165837	653	426409	25.5539	.00153139
604	364816	24.5764	.00165563	654	427716	25.5734	.00152905
605	366025	24.5967	.00165289	655	429025	25.5930	.00152672
606	367236	24.6171	.00165017	656	430336	25.6125	.00152439
607	368449	24.6374	.00164745	657	431649	25.6320	.00152207
608	369664	24.6577	.00164474	658	432964	25.6515	.00151976
609	370881	24.6779	.00164204	659	434281	25.6710	.00151745
610	372100	24.6982	.00163934	660	435600	25.6905	.00151515
611	373321	24.7184	.00163666	661	436921	25.7099	.00151286
612	374544	24.7386	.00163399	662	438244	25.7294	.00151057
613	375769	24.7588	.00163132	663	439569	25.7488	.00150830
614	376996	24.7790	.00162866	664	440896	25.7682	.00150602
615	378225	24.7992	.00162602	665	442225	25.7876	.00150376
616	379456	24.8193	.00162338	666	443556	25.8070	.00150150
617	380689	24.8395	.00162075	667	444889	25.8263	.00149925
618	381924	24.8596	.00161812	668	446224	25.8457	.00149701
619	383161	24.8797	.00161551	669	447561	25.8650	.00149477
620	384400	24.8998	.00161290	670	448900	25.8844	.00149254
621	385641	24.9199	.00161031	671	450241	25.9037	.00149031
622	386884	24.9399	.00160772	672	451584	25.9230	.00148810
623	388129	24.9600	.00160514	673	452929	25.9422	.00148588
624	389376	24.9800	.00160256	674	454276	25.9615	.00148368
625	390625	25.0000	.00160000	675	455625	25.9808	.00148148
626	391876	25.0200	.00159744	676	456976	26.0000	.00147929
627	393129	25.0400	.00159490	677	458329	26.0192	.00147710
628	394384	25.0599	.00159236	678	459684	26.0384	.00147493
629	395641	25.0799	.00158983	679	461041	26.0576	.00147275
630	396900	25.0998	.00158730	680	462400	26.0768	.00147059
631	398161	25.1197	.00158479	681	463761	26.0960	.00146843
632	399424	25.1396	.00158228	682	465124	26.1151	.00146628
633	400689	25.1595	.00157978	683	466489	26.1343	.00146413
634	401956	25.1794	.00157729	684	467856	26.1534	.00146199
635	403225	25.1992	.00157480	685	469225	26.1725	.00145985
636	404496	25.2190	.00157233	686	470596	26.1916	.00145773
637	405769	25.2389	.00156986	687	471969	26.2107	.00145560
638	407044	25.2587	.00156740	688	473344	26.2298	.00145349
639	408321	25.2784	.00156495	689	474721	26.2488	.00145138
640	409600	25.2982	.00156250	690	476100	26.2679	.00144928

N	N^2	$\sqrt{N}$	$1/N$	N	N^2	$\sqrt{N}$	$1/N$
691	477481	26.2869	.00144718	741	549081	27.2213	.00134953
692	478864	26.3059	.00144509	742	550564	27.2397	.00134771
693	480249	26.3249	.00144300	743	552049	27.2580	.00134590
694	481636	26.3439	.00144092	744	553536	27.2764	.00134409
695	483025	26.3629	.00143885	745	555025	27.2947	.00134228
696	484416	26.3818	.00143678	746	556516	27.3130	.00134048
697	485809	26.4008	.00143472	747	558009	27.3313	.00133869
698	487204	26.4197	.00143266	748	559504	27.3496	.00133690
699	488601	26.4386	.00143062	749	561001	27.3679	.00133511
700	490000	26.4575	.00142857	750	562500	27.3861	.00133333
701	491401	26.4764	.00142653	751	564001	27.4044	.00133156
702	492804	26.4953	.00142450	752	565504	27.4226	.00132979
703	494209	26.5141	.00142248	753	567009	27.4408	.00132802
704	495616	26.5330	.00142045	754	568516	27.4591	.00132626
705	497025	26.5518	.00141844	755	570025	27.4773	.00132450
706	498436	26.5707	.00141643	756	571536	27.4955	.00132275
707	499849	26.5895	.00141443	757	573049	27.5136	.00132100
708	501264	26.6083	.00141243	758	574564	27.5318	.00131926
709	502681	26.6271	.00141044	759	576081	27.5500	.00131752
710	504100	26.6458	.00140845	760	577600	27.5681	.00131579
711	505521	26.6646	.00140647	761	579121	27.5862	.00131406
712	506944	26.6833	.00140449	762	580644	27.6043	.00131234
713	508369	26.7021	.00140252	763	582169	27.6225	.00131062
714	509796	26.7208	.00140056	764	583696	27.6405	.00130890
715	511225	26.7395	.00139860	765	585225	27.6586	.00130719
716	512656	26.7582	.00139665	766	586756	27.6767	.00130548
717	514089	26.7769	.00139470	767	588289	27.6948	.00130378
718	515524	26.7955	.00139276	768	589824	27.7128	.00130208
719	516961	26.8142	.00139082	769	591361	27.7308	.00130039
720	518400	26.8328	.00138889	770	592900	27.7489	.00129870
721	519841	26.8514	.00138696	771	594441	27.7669	.00129702
722	521284	26.8701	.00138504	772	595984	27.7849	.00129534
723	522729	26.8887	.00138313	773	597529	27.8029	.00129366
724	524176	26.9072	.00138122	774	599076	27.8209	.00129199
725	525625	26.9258	.00137931	775	600625	27.8388	.00129032
726	527076	26.9444	.00137741	776	602176	27.8568	.00128866
727	528529	26.9629	.00137552	777	603729	27.8747	.00128700
728	529984	26.9815	.00137363	778	605284	27.8927	.00128535
729	531441	27.0000	.00137174	779	606841	27.9106	.00128370
730	532900	27.0185	.00136986	780	608400	27.9285	.00128205
731	534361	27.0370	.00136799	781	609961	27.9464	.00128041
732	535824	27.0555	.00136612	782	611524	27.9643	.00127877
733	537289	27.0740	.00136426	783	613089	27.9821	.00127714
734	538756	27.0924	.00136240	784	614656	28.0000	.00127551
735	540225	27.1109	.00136054	785	616225	28.0179	.00127389
736	541696	27.1293	.00135870	786	617796	28.0357	.00127226
737	543169	27.1477	.00135685	787	619369	28.0535	.00127065
738	544644	27.1662	.00135501	788	620944	28.0713	.00126904
739	546121	27.1846	.00135318	789	622521	28.0891	.00126743
740	547600	27.2029	.00135135	790	624100	28.1069	.00126582

(*Continued*)

N	N^2	$\sqrt{N}$	$1/N$	N	N^2	$\sqrt{N}$	$1/N$
791	625681	28.1247	.00126422	836	698896	28.9137	.00119617
792	627264	28.1425	.00126263	837	700569	28.9310	.00119474
793	628849	28.1603	.00126103	838	702244	28.9482	.00119332
794	630436	28.1780	.00125945	839	703921	28.9655	.00119190
795	632025	28.1957	.00125786	840	705600	28.9828	.00119048
796	633616	28.2135	.00125628	841	707281	29.0000	.00118906
797	635209	28.2312	.00125471	842	708964	29.0172	.00118765
798	636804	28.2489	.00125313	843	710649	29.0345	.00118624
799	638401	28.2666	.00125156	844	712336	29.0517	.00118483
800	640000	28.2843	.00125000	845	714025	29.0689	.00118343
801	641601	28.3019	.00124844	846	715716	29.0861	.00118203
802	643204	28.3196	.00124688	847	717409	29.1033	.00118064
803	644809	28.3373	.00124533	848	719104	29.1204	.00117925
804	646416	28.3549	.00124378	849	720801	29.1376	.00117786
805	648025	28.3725	.00124224	850	722500	29.1548	.00117647
806	649636	28.3901	.00124069	851	724201	29.1719	.00117509
807	651249	28.4077	.00123916	852	725904	29.1890	.00117371
808	652864	28.4253	.00123762	853	727609	29.2062	.00117233
809	654481	28.4429	.00123609	854	729316	29.2233	.00117096
810	656100	28.4605	.00123457	855	731025	29.2404	.00116959
811	657721	28.4781	.00123305	856	732736	29.2575	.00116822
812	659344	28.4956	.00123153	857	734449	29.2746	.00116686
813	660969	28.5132	.00123001	858	736164	29.2916	.00116550
814	662596	28.5307	.00122850	859	737881	29.3087	.00116414
815	664225	28.5482	.00122699	860	739600	29.3258	.00116279
816	665856	28.5657	.00122549	861	741321	29.3428	.00116144
817	667489	28.5832	.00122399	862	743044	29.3598	.00116009
818	669124	28.6007	.00122249	863	744769	29.3769	.00115875
819	670761	28.6182	.00122100	864	746496	29.3939	.00115741
820	672400	28.6356	.00121951	865	748225	29.4109	.00115607
821	674041	28.6531	.00121803	866	749956	29.4279	.00115473
822	675684	28.6705	.00121655	867	751689	29.4449	.00115340
823	677329	28.6880	.00121507	868	753424	29.4618	.00115207
824	678976	28.7054	.00121359	869	755161	29.4788	.00115075
825	680625	28.7228	.00121212	870	756900	29.4958	.00114943
826	682276	28.7402	.00121065	871	758641	29.5127	.00114811
827	683929	28.7576	.00120919	872	760384	29.5296	.00114679
828	685584	28.7750	.00120773	873	762129	29.5466	.00114548
829	687241	28.7924	.00120627	874	763876	29.5635	.00114416
830	688900	28.8097	.00120482	875	765625	29.5804	.00114286
831	690561	28.8271	.00120337	876	767376	29.5973	.00114155
832	692224	28.8444	.00120192	877	769129	29.6142	.00114025
833	693889	28.8617	.00120048	878	770884	29.6311	.00113895
834	695556	28.8791	.00119904	879	772641	29.6479	.00113766
835	697225	28.8964	.00119760	880	774400	29.6648	.00113636

N	N^2	$\sqrt{N}$	$1/N$	N	N^2	$\sqrt{N}$	$1/N$
881	776161	29.6816	.00113507	926	857476	30.4302	.00107991
882	777924	29.6985	.00113379	927	859329	30.4467	.00107875
883	779689	29.7153	.00113250	928	861184	30.4631	.00107759
884	781456	29.7321	.00113122	929	863041	30.4795	.00107643
885	783225	29.7489	.00112994	930	864900	30.4959	.00107527
886	784996	29.7658	.00112867	931	866761	30.5123	.00107411
887	786769	29.7825	.00112740	932	868624	30.5287	.00107296
888	788544	29.7993	.00112613	933	870489	30.5450	.00107181
889	790321	29.8161	.00112486	934	872356	30.5614	.00107066
890	792100	29.8329	.00112360	935	874225	30.5778	.00106952
891	793881	29.8496	.00112233	936	876096	30.5941	.00106838
892	795664	29.8664	.00112108	937	877969	30.6105	.00106724
893	797449	29.8831	.00111982	938	879844	30.6268	.00106610
894	799236	29.8998	.00111857	939	881721	30.6431	.00106496
895	801025	29.9166	.00111732	940	883600	30.6594	.00106383
896	802816	29.9333	.00111607	941	885481	30.6757	.00106270
897	804609	29.9500	.00111483	942	887364	30.6920	.00106157
898	806404	29.9666	.00111359	943	889249	30.7083	.00106045
899	808201	29.9833	.00111235	944	891136	30.7246	.00105932
900	810000	30.0000	.00111111	945	893025	30.7409	.00105820
901	811801	30.0167	.00110988	946	894916	30.7571	.00105708
902	813604	30.0333	.00110865	947	896809	30.7734	.00105597
903	815409	30.0500	.00110742	948	898704	30.7896	.00105485
904	817216	30.0666	.00110619	949	900601	30.8058	.00105374
905	819025	30.0832	.00110497	950	902500	30.8221	.00105263
906	820836	30.0998	.00110375	951	904401	30.8383	.00105152
907	822649	30.1164	.00110254	952	906304	30.8545	.00105042
908	824464	30.1330	.00110132	953	908209	30.8707	.00104932
909	826281	30.1496	.00110011	954	910116	30.8869	.00104822
910	828100	30.1662	.00109890	955	912025	30.9031	.00104712
911	829921	30.1828	.00109769	956	913936	30.9192	.00104603
912	831744	30.1993	.00109649	957	915849	30.9354	.00104493
913	833569	30.2159	.00109529	958	917764	30.9516	.00104384
914	835396	30.2324	.00109409	959	919681	30.9677	.00104275
915	837225	30.2490	.00109290	960	921600	30.9839	.00104167
916	839056	30.2655	.00109170	961	923521	31.0000	.00104058
917	840889	30.2820	.00109051	962	925444	31.0161	.00103950
918	842724	30.2985	.00108932	963	927369	31.0322	.00103842
919	844561	30.3150	.00108814	964	929296	31.0483	.00103734
920	846400	30.3315	.00108696	965	931225	31.0644	.00103627
921	848241	30.3480	.00108578	966	933156	31.0805	.00103520
922	850084	30.3645	.00108460	967	935089	31.0966	.00103413
923	851929	30.3809	.00108342	968	937024	31.1127	.00103306
924	853776	30.3974	.00108225	969	938961	31.1288	.00103199
925	855625	30.4138	.00108108	970	940900	31.1448	.00103093

(*Continued*)

N	N^2	$\sqrt{N}$	$1/N$	N	N^2	$\sqrt{N}$	$1/N$
971	942841	31.1609	.00102987	986	972196	31.4006	.00101420
972	944784	31.1769	.00102881	987	974169	31.4166	.00101317
973	946729	31.1929	.00102775	988	976144	31.4325	.00101215
974	948676	31.2090	.00102669	989	978121	31.4484	.00101112
975	950625	31.2250	.00102564	990	980100	31.4643	.00101010
976	952576	31.2410	.00102459	991	982081	31.4802	.00100908
977	954529	31.2570	.00102354	992	984064	31.4960	.00100806
978	956484	31.2730	.00102249	993	986049	31.5119	.00100705
979	958441	31.2890	.00102145	994	988036	31.5278	.00100604
980	960400	31.3050	.00102041	995	990025	31.5436	.00100503
981	962361	31.3209	.00101937	996	992016	31.5595	.00100402
982	964324	31.3369	.00101833	997	994009	31.5753	.00100301
983	966289	31.3528	.00101729	998	996004	31.5911	.00100200
984	968256	31.3688	.00101626	999	998001	31.6070	.00100100
985	970225	31.3847	.00101523	1000	1000000	31.6228	.00100000

TABLE II Areas and ordinates of the normal curve in terms of $Z = (X - \mu)/\sigma$

(1) Z	(2) A Area from Mean to Z	(3) B Area in Larger Portion	(4) C Area in Smaller Portion	(5) y Ordinate at Z
0.00	.0000	.5000	.5000	.3989
0.01	.0040	.5040	.4960	.3989
0.02	.0080	.5080	.4920	.3989
0.03	.0120	.5120	.4880	.3988
0.04	.0160	.5160	.4840	.3986
0.05	.0199	.5199	.4801	.3984
0.06	.0239	.5239	.4761	.3982
0.07	.0279	.5279	.4721	.3980
0.08	.0319	.5319	.4681	.3977
0.09	.0359	.5359	.4641	.3973
0.10	.0398	.5398	.4602	.3970
0.11	.0438	.5438	.4562	.3965
0.12	.0478	.5478	.4522	.3961
0.13	.0517	.5517	.4483	.3956
0.14	.0557	.5557	.4443	.3951
0.15	.0596	.5596	.4404	.3945
0.16	.0636	.5636	.4364	.3939
0.17	.0675	.5675	.4325	.3932
0.18	.0714	.5714	.4286	.3925
0.19	.0753	.5753	.4247	.3918
0.20	.0793	.5793	.4207	.3910
0.21	.0832	.5832	.4168	.3902
0.22	.0871	.5871	.4129	.3894
0.23	.0910	.5910	.4090	.3885
0.24	.0948	.5948	.4052	.3876
0.25	.0987	.5987	.4013	.3867
0.26	.1026	.6026	.3974	.3857
0.27	.1064	.6064	.3936	.3847
0.28	.1103	.6103	.3897	.3836
0.29	.1141	.6141	.3859	.3825
0.30	.1179	.6179	.3821	.3814
0.31	.1217	.6217	.3783	.3802
0.32	.1255	.6255	.3745	.3790
0.33	.1293	.6293	.3707	.3778
0.34	.1331	.6331	.3669	.3765
0.35	.1368	.6368	.3632	.3752
0.36	.1406	.6406	.3594	.3739
0.37	.1443	.6443	.3557	.3725
0.36	.1480	.6480	.3520	.3712
0.39	.1517	.6517	.3483	.3697
0.40	.1554	.6554	.3446	.3683
0.41	.1591	.6591	.3409	.3668
0.42	.1628	.6628	.3372	.3653
0.43	.1664	.6664	.3336	.3637
0.44	.1700	.6700	.3300	.3621

(*Continued*)

(1) Z	(2) A Area from Mean to Z	(3) B Area in Larger Portion	(4) C Area in Smaller Portion	(5) y Ordinate at Z
0.45	.1736	.6736	.3264	.3605
0.46	.1772	.6772	.3228	.3589
0.47	.1808	.6808	.3192	.3572
0.48	.1844	.6844	.3156	.3555
0.49	.1879	.6879	.3121	.3538
0.50	.1915	.6915	.3085	.3521
0.51	.1950	.6950	.3050	.3503
0.52	.1985	.6985	.3015	.3485
0.53	.2019	.7019	.2981	.3467
0.54	.2054	.7054	.2946	.3448
0.55	.2088	.7088	.2912	.3429
0.56	.2123	.7123	.2877	.3410
0.57	.2157	.7157	.2843	.3391
0.58	.2190	.7190	.2810	.3372
0.59	.2224	.7224	.2776	.3352
0.60	.2257	.7257	.2743	.3332
0.61	.2291	.7291	.2709	.3312
0.62	.2324	.7324	.2676	.3292
0.63	.2357	.7357	.2643	.3271
0.64	.2389	.7389	.2611	.3251
0.65	.2422	.7422	.2578	.3230
0.66	.2454	.7454	.2546	.3209
0.67	.2486	.7486	.2514	.3187
0.68	.2517	.7517	.2483	.3166
0.69	.2549	.7549	.2451	.3144
0.70	.2580	.7580	.2420	.3123
0.71	.2611	.7611	.2389	.3101
0.72	.2642	.7642	.2358	.3079
0.73	.2673	.7673	.2327	.3056
0.74	.2704	.7704	.2296	.3034
0.75	.2734	.7734	.2266	.3011
0.76	.2764	.7764	.2236	.2989
0.77	.2794	.7794	.2206	.2966
0.78	.2823	.7823	.2177	.2943
0.79	.2852	.7852	.2148	.2920
0.80	.2881	.7881	.2119	.2897
0.81	.2910	.7910	.2090	.2874
0.82	.2939	.7939	.2061	.2850
0.83	.2967	.7967	.2033	.2827
0.84	.2995	.7995	.2005	.2803
0.85	.3023	.8023	.1977	.2780
0.86	.3051	.8051	.1949	.2756
0.87	.3078	.8078	.1922	.2732
0.88	.3106	.8106	.1894	.2709
0.89	.3133	.8133	.1867	.2685

(1) Z	(2) A Area from Mean to Z	(3) B Area in Larger Portion	(4) C Area in Smaller Portion	(5) y Ordinate at Z
0.90	.3159	.8159	.1841	.2661
0.91	.3186	.8186	.1814	.2637
0.92	.3212	.8212	.1788	.2613
0.93	.3238	.8238	.1762	.2589
0.94	.3264	.8264	.1736	.2565
0.95	.3289	.8289	.1711	.2541
0.96	.3315	.8315	.1685	.2516
0.97	.3340	.8340	.1660	.2492
0.98	.3365	.8365	.1635	.2468
0.99	.3389	.8389	.1611	.2444
1.00	.3413	.8413	.1587	.2420
1.01	.3438	.8438	.1562	.2396
1.02	.3461	.8461	.1539	.2371
1.03	.3485	.8485	.1515	.2347
1.04	.3508	.8508	.1492	.2323
1.05	.3531	.8531	.1469	.2299
1.06	.3554	.8554	.1446	.2275
1.07	.3577	.8577	.1423	.2251
1.08	.3599	.8599	.1401	.2227
1.09	.3621	.8621	.1379	.2203
1.10	.3643	.8643	.1357	.2179
1.11	.3665	.8665	.1335	.2155
1.12	.3686	.8686	.1314	.2131
1.13	.3708	.8708	.1292	.2107
1.14	.3729	.8729	.1271	.2083
1.15	.3749	.8749	.1251	.2059
1.16	.3770	.8770	.1230	.2036
1.17	.3790	.8790	.1210	.2012
1.18	.3810	.8810	.1190	.1989
1.19	.3830	.8830	.1170	.1965
1.20	.3849	.8849	.1151	.1942
1.21	.3869	.8869	.1131	.1919
1.22	.3888	.8888	.1112	.1895
1.23	.3907	.8907	.1093	.1872
1.24	.3925	.8925	.1075	.1849
1.25	.3944	.8944	.1056	.1826
1.26	.3962	.8962	.1038	.1804
1.27	.3980	.8980	.1020	.1781
1.28	.3997	.8997	.1003	.1758
1.29	.4015	.9015	.0985	.1736
1.30	.4032	.9032	.0968	.1714
1.31	.4049	.9049	.0951	.1691
1.32	.4066	.9066	.0934	.1669
1.33	.4082	.9082	.0918	.1647
1.34	.4099	.9099	.0901	.1626

(*Continued*)

(1) Z	(2) A Area from Mean to Z	(3) B Area in Larger Portion	(4) C Area in Smaller Portion	(5) y Ordinate at Z
1.35	.4115	.9115	.0885	.1604
1.36	.4131	.9131	.0869	.1582
1.37	.4147	.9147	.0853	.1561
1.38	.4162	.9162	.0838	.1539
1.39	.4177	.9177	.0823	.1518
1.40	.4192	.9192	.0808	.1497
1.41	.4207	.9207	.0793	.1476
1.42	.4222	.9222	.0778	.1456
1.43	.4236	.9236	.0764	.1435
1.44	.4251	.9251	.0749	.1415
1.45	.4265	.9265	.0735	.1394
1.46	.4279	.9279	.0721	.1374
1.47	.4292	.9292	.0708	.1354
1.48	.4306	.9306	.0694	.1334
1.49	.4319	.9319	.0681	.1315
1.50	.4332	.9332	.0668	.1295
1.51	.4345	.9345	.0655	.1276
1.52	.4357	.9357	.0643	.1257
1.53	.4370	.9370	.0630	.1238
1.54	.4382	.9382	.0618	.1219
1.55	.4394	.9394	.0606	.1200
1.56	.4406	.9406	.0594	.1182
1.57	.4418	.9418	.0582	.1163
1.58	.4429	.9429	.0571	.1145
1.59	.4441	.9441	.0559	.1127
1.60	.4452	.9452	.0548	.1109
1.61	.4463	.9463	.0537	.1092
1.62	.4474	.9474	.0526	.1074
1.63	.4484	.9484	.0516	.1057
1.64	.4495	.9495	.0505	.1040
1.65	.4505	.9505	.0495	.1023
1.66	.4515	.9515	.0485	.1006
1.67	.4525	.9525	.0475	.0989
1.68	.4535	.9535	.0465	.0973
1.69	.4545	.9545	.0455	.0957
1.70	.4554	.9554	.0446	.0940
1.71	.4564	.9564	.0436	.0925
1.72	.4573	.9573	.0427	.0909
1.73	.4582	.9582	.0418	.0893
1.74	.4591	.9591	.0409	.0878
1.75	.4599	.9599	.0401	.0863
1.76	.4608	.9608	.0392	.0848
1.77	.4616	.9616	.0384	.0833
1.78	.4625	.9625	.0375	.0818
1.79	.4633	.9633	.0367	.0804

(1) Z	(2) A Area from Mean to Z	(3) B Area in Larger Portion	(4) C Area in Smaller Portion	(5) y Ordinate at Z
1.80	.4641	.9641	.0359	.0790
1.81	.4649	.9649	.0351	.0775
1.82	.4656	.9656	.0344	.0761
1.83	.4664	.9664	.0336	.0748
1.84	.4671	.9671	.0329	.0734
1.85	.4678	.9678	.0322	.0721
1.86	.4686	.9686	.0314	.0707
1.87	.4693	.9693	.0307	.0694
1.88	.4699	.9699	.0301	.0681
1.89	.4706	.9706	.0294	.0669
1.90	.4713	.9713	.0287	.0656
1.91	.4719	.9719	.0281	.0644
1.92	.4726	.9726	.0274	.0632
1.93	.4732	.9732	.0268	.0620
1.94	.4738	.9738	.0262	.0608
1.95	.4744	.9744	.0256	.0596
1.96	.4750	.9750	.0250	.0584
1.97	.4756	.9756	.0244	.0573
1.98	.4761	.9761	.0239	.0562
1.99	.4767	.9767	.0233	.0551
2.00	.4772	.9772	.0228	.0540
2.01	.4778	.9778	.0222	.0529
2.02	.4783	.9783	.0217	.0519
2.03	.4788	.9788	.0212	.0508
2.04	.4793	.9793	.0207	.0498
2.05	.4798	.9798	.0202	.0488
2.06	.4803	.9803	.0197	.0478
2.07	.4808	.9808	.0192	.0468
2.08	.4812	.9812	.0188	.0459
2.09	.4817	.9817	.0183	.0449
2.10	.4821	.9821	.0179	.0440
2.11	.4826	.9826	.0174	.0431
2.12	.4830	.9830	.0170	.0422
2.13	.4834	.9834	.0166	.0413
2.14	.4838	.9838	.0162	.0404
2.15	.4842	.9842	.0158	.0396
2.16	.4846	.9846	.0154	.0387
2.17	.4850	.9850	.0150	.0379
2.18	.4854	.9854	.0146	.0371
2.19	.4857	.9857	.0143	.0363
2.20	.4861	.9861	.0139	.0355
2.21	.4864	.9864	.0136	.0347
2.22	.4868	.9868	.0132	.0339
2.23	.4871	.9871	.0129	.0332
2.24	.4875	.9875	.0125	.0325

(*Continued*)

(1) Z	(2) A Area from Mean to Z	(3) B Area in Larger Portion	(4) C Area in Smaller Portion	(5) y Ordinate at Z
2.25	.4878	.9878	.0122	.0317
2.26	.4881	.9881	.0119	.0310
2.27	.4884	.9884	.0116	.0303
2.28	.4887	.9887	.0113	.0297
2.29	.4890	.9890	.0110	.0290
2.30	.4893	.9893	.0107	.0283
2.31	.4896	.9896	.0104	.0277
2.32	.4898	.9898	.0102	.0270
2.33	.4901	.9901	.0099	.0264
2.34	.4904	.9904	.0096	.0258
2.35	.4906	.9906	.0094	.0252
2.36	.4909	.9909	.0091	.0246
2.37	.4911	.9911	.0089	.0241
2.38	.4913	.9913	.0087	.0235
2.39	.4916	.9916	.0084	.0229
2.40	.4918	.9918	.0082	.0224
2.41	.4920	.9920	.0080	.0219
2.42	.4922	.9922	.0078	.0213
2.43	.4925	.9925	.0075	.0208
2.44	.4927	.9927	.0073	.0203
2.45	.4929	.9929	.0071	.0198
2.46	.4931	.9931	.0069	.0194
2.47	.4932	.9932	.0068	.0189
2.48	.4934	.9934	.0066	.0184
2.49	.4936	.9936	.0064	.0180
2.50	.4938	.9938	.0062	.0175
2.51	.4940	.9940	.0060	.0171
2.52	.4941	.9941	.0059	.0167
2.53	.4943	.9943	.0057	.0163
2.54	.4945	.9945	.0055	.0158
2.55	.4946	.9946	.0054	.0154
2.56	.4948	.9948	.0052	.0151
2.57	.4949	.9949	.0051	.0147
2.58	.4951	.9951	.0049	.0143
2.59	.4952	.9952	.0048	.0139
2.60	.4953	.9953	.0047	.0136
2.61	.4955	.9955	.0045	.0132
2.62	.4956	.9956	.0044	.0129
3.63	.4957	.9957	.0043	.0126
2.64	.4959	.9959	.0041	.0122
2.65	.4960	.9960	.0040	.0119
2.66	.4961	.9961	.0039	.0116
2.67	.4962	.9962	.0038	.0113
2.68	.4963	.9963	.0037	.0110
2.69	.4964	.9964	.0036	.0107

(1) Z	(2) A Area from Mean to Z	(3) B Area in Larger Portion	(4) C Area in Smaller Portion	(5) y Ordinate at Z
2.70	.4965	.9965	.0035	.0104
2.71	.4966	.9966	.0034	.0101
2.72	.4967	.9967	.0033	.0099
2.73	.4968	.9968	.0032	.0096
2.74	.4969	.9969	.0031	.0093
2.75	.4970	.9970	.0030	.0091
2.76	.4971	.9971	.0029	.0088
2.77	.4972	.9972	.0028	.0086
2.78	.4973	.9973	.0027	.0084
2.79	.4974	.9974	.0026	.0081
2.80	.4974	.9974	.0026	.0079
2.81	.4975	.9975	.0025	.0077
2.82	.4976	.9976	.0024	.0075
2.83	.4977	.9977	.0023	.0073
2.84	.4977	.9977	.0023	.0071
2.85	.4978	.9978	.0022	.0069
2.86	.4979	.9979	.0021	.0067
2.87	.4979	.9979	.0021	.0065
2.88	.4980	.9980	.0020	.0063
2.89	.4981	.9981	.0019	.0061
2.90	.4981	.9981	.0019	.0060
2.91	.4982	.9982	.0018	.0058
2.92	.4982	.9982	.0018	.0056
2.93	.4983	.9983	.0017	.0055
2.94	.4984	.9984	.0016	.0053
2.95	.4984	.9984	.0016	.0051
2.96	.4985	.9985	.0015	.0050
2.97	.4985	.9985	.0015	.0048
2.98	.4986	.9986	.0014	.0047
2.99	.4986	.9986	.0014	.0046
3.00	.4987	.9987	.0013	.0044
3.01	.4987	.9987	.0013	.0043
3.02	.4987	.9987	.0013	.0042
3.03	.4988	.9988	.0012	.0040
3.04	.4988	.9988	.0012	.0039
3.05	.4989	.9989	.0011	.0038
3.06	.4989	.9989	.0011	.0037
3.07	.4989	.9989	.0011	.0036
3.08	.4990	.9990	.0010	.0035
3.09	.4990	.9990	.0010	.0034
3.10	.4990	.9990	.0010	.0033
3.11	.4991	.9991	.0009	.0032
3.12	.4991	.9991	.0009	.0031
3.13	.4991	.9991	.0009	.0030
3.14	.4992	.9992	.0008	.0029

(*Continued*)

(1) Z	(2) A Area from Mean to Z	(3) B Area in Larger Portion	(4) C Area in Smaller Portion	(5) y Ordinate at Z
3.15	.4992	.9992	.0008	.0028
3.16	.4992	.9992	.0008	.0027
3.17	.4992	.9992	.0008	.0026
3.18	.4993	.9993	.0007	.0025
3.19	.4993	.9993	.0007	.0025
3.20	.4993	.9993	.0007	.0024
3.21	.4993	.9993	.0007	.0023
3.22	.4994	.9994	.0006	.0022
3.23	.4994	.9994	.0006	.0022
3.24	.4994	.9994	.0006	.0021
3.30	.4995	.9995	.0005	.0017
3.40	.4997	.9997	.0003	.0012
3.50	.4998	.9998	.0002	.0009
3.60	.4998	.9998	.0002	.0006
3.70	.4999	.9999	.0001	.0004

TABLE III Table of χ^2

The probabilities given by the column headings are those for obtaining χ^2 equal to or greater than the tabled value when a null hypothesis is true and when χ^2 has the degrees of freedom given by the column at the left.

Degrees of freedom *df*	*P* = .99	.98	.95	.90	.80	.70	.50	.30	.20	.10	.05	.02	.01
1	.000157	.000628	.00393	.0158	.0642	.148	.455	1.074	1.642	2.706	3.841	5.412	6.635
2	.0201	.0404	.103	.211	.446	.713	1.386	2.408	3.219	4.605	5.991	7.824	9.210
3	.115	.185	.352	.584	1.005	1.424	2.366	3.665	4.642	6.251	7.815	9.837	11.341
4	.297	.429	.711	1.064	1.649	2.195	3.357	4.878	5.989	7.779	9.488	11.668	13.277
5	.554	.752	1.145	1.610	2.343	3.000	4.351	6.064	7.289	9.236	11.070	13.388	15.086
6	.872	1.134	1.635	2.204	3.070	3.828	5.348	7.231	8.558	10.645	12.592	15.033	16.812
7	1.239	1.564	2.167	2.833	3.822	4.671	6.346	8.383	9.803	12.017	14.067	16.622	18.475
8	1.646	2.032	2.733	3.490	4.594	5.527	7.344	9.524	11.030	13.362	15.507	18.168	20.090
9	2.088	2.532	3.325	4.168	5.380	6.393	8.343	10.656	12.242	14.684	16.919	19.679	21.666
10	2.558	3.059	3.940	4.865	6.179	7.267	9.342	11.781	13.442	15.987	18.307	21.161	23.209
11	3.053	3.609	4.575	5.578	6.989	8.148	10.341	12.899	14.631	17.275	19.675	22.618	24.725
12	3.571	4.178	5.226	6.304	7.807	9.034	11.340	14.011	15.812	18.549	21.026	24.054	26.217
13	4.107	4.765	5.892	7.042	8.634	9.926	12.340	15.119	16.985	19.812	22.362	25.472	27.688
14	4.660	5.368	6.571	7.790	9.467	10.821	13.339	16.222	18.151	21.064	23.685	26.873	29.141
15	5.229	5.985	7.261	8.547	10.307	11.721	14.339	17.322	19.311	22.307	24.996	28.259	30.578
16	5.812	6.614	7.962	9.312	11.152	12.624	15.338	18.418	20.465	23.542	26.296	29.633	32.000
17	6.408	7.255	8.672	10.085	12.002	13.531	16.338	19.511	21.615	24.769	27.587	30.995	33.409
18	7.015	7.906	9.390	10.865	12.857	14.440	17.338	20.601	22.760	25.989	28.869	32.346	34.805
19	7.633	8.567	10.117	11.651	13.716	15.352	18.338	21.689	23.900	27.204	30.144	33.687	36.191
20	8.260	9.237	10.851	12.443	14.578	16.266	19.337	22.775	25.038	28.412	31.410	35.020	37.566
21	8.897	9.915	11.591	13.240	15.445	17.182	20.337	23.858	26.171	29.615	32.671	36.343	38.932
22	9.542	10.600	12.338	14.041	16.314	18.101	21.337	24.939	27.301	30.813	33.924	37.659	40.289
23	10.196	11.293	13.091	14.848	17.187	19.021	22.337	26.018	28.429	32.007	35.172	38.968	41.638
24	10.856	11.992	13.848	15.659	18.062	19.943	23.337	27.096	29.553	33.196	36.415	40.270	42.980
25	11.524	12.697	14.611	16.473	18.940	20.867	24.337	28.172	30.675	34.382	37.652	41.566	44.314
26	12.198	13.409	15.379	17.292	19.820	21.792	25.336	29.246	31.795	35.563	38.885	42.856	45.642
27	12.879	14.125	16.151	18.114	20.703	22.719	26.336	30.319	32.912	36.741	40.113	44.140	46.963
28	13.565	14.847	16.928	18.939	21.588	23.647	27.336	31.391	34.027	37.916	41.337	45.419	48.278
29	14.256	15.574	17.708	19.768	22.475	24.577	28.336	32.461	35.139	39.087	42.557	46.693	49.588
30	14.953	16.306	18.493	20.599	23.364	25.508	29.336	33.530	36.250	40.256	43.773	47.962	50.892

SOURCE: Reprinted from Table III of R. A. Fisher, *Statistical Methods for Research Workers* (14th ed.).

TABLE IV Table of t

The probabilities given by the column headings are for a one-sided test, assuming a null hypothesis to be true. For example, with 30 d.f., we have $P(t \geq 2.042) = .025$. For a two-sided test, we have $P(t \geq 2.042) + P(t \leq -2.042) = .025 + .025 = .05$.

df \ P	.25	.10	.05	.025	.01	.005	.0025	.001
1	1.000	3.078	6.314	12.706	31.821	63.657	127.321	318.309
2	.816	1.886	2.920	4.303	6.965	9.925	14.089	22.327
3	.765	1.638	2.353	3.182	4.541	5.841	7.453	10.214
4	.741	1.533	2.132	2.776	3.747	4.604	5.598	7.173
5	.727	1.476	2.015	2.571	3.365	4.032	4.773	5.893
6	.718	1.440	1.943	2.447	3.143	3.707	4.317	5.208
7	.711	1.415	1.895	2.365	2.998	3.499	4.029	4.785
8	.706	1.397	1.860	2.306	2.896	3.355	3.833	4.501
9	.703	1.383	1.833	2.262	2.821	3.250	3.690	4.297
10	.700	1.372	1.812	2.228	2.764	3.169	3.581	4.144
11	.697	1.363	1.796	2.201	2.718	3.106	3.497	4.025
12	.695	1.356	1.782	2.179	2.681	3.055	3.428	3.930
13	.694	1.350	1.771	2.160	2.650	3.012	3.372	3.852
14	.692	1.345	1.761	2.145	2.624	2.977	3.326	3.787
15	.691	1.341	1.753	2.131	2.602	2.947	3.286	3.733
16	.690	1.337	1.746	2.120	2.583	2.921	3.252	3.686
17	.689	1.333	1.740	2.110	2.567	2.898	3.223	3.646
18	.688	1.330	1.734	2.101	2.552	2.878	3.197	3.610
19	.688	1.328	1.729	2.093	2.539	2.861	3.174	3.579
20	.687	1.325	1.725	2.086	2.528	2.845	3.153	3.552
21	.686	1.323	1.721	2.080	2.518	2.831	3.135	3.527
22	.686	1.321	1.717	2.074	2.508	2.819	3.119	3.505
23	.685	1.319	1.714	2.069	2.500	2.807	3.104	3.485
24	.685	1.318	1.711	2.064	2.492	2.797	3.090	3.467
25	.684	1.316	1.708	2.060	2.485	2.787	3.078	3.450
26	.684	1.315	1.706	2.056	2.479	2.779	3.067	3.435
27	.684	1.314	1.703	2.052	2.473	2.771	3.057	3.421
28	.683	1.313	1.701	2.048	2.467	2.763	3.047	3.408
29	.683	1.311	1.699	2.045	2.462	2.756	3.038	3.396
30	.683	1.310	1.697	2.042	2.457	2.750	3.030	3.385
35	.682	1.306	1.690	2.030	2.438	2.724	2.996	3.340
40	.681	1.303	1.684	2.021	2.423	2.704	2.971	3.307
45	.680	1.301	1.679	2.014	2.412	2.690	2.952	3.281
50	.679	1.299	1.676	2.009	2.403	2.678	2.937	3.261
55	.679	1.297	1.673	2.004	2.396	2.668	2.925	3.245
60	.679	1.296	1.671	2.000	2.390	2.660	2.915	3.232
70	.678	1.294	1.667	1.994	2.381	2.648	2.899	3.211
80	.678	1.292	1.664	1.990	2.374	2.639	2.887	3.195
90	.677	1.291	1.662	1.987	2.368	2.632	2.878	3.183
100	.677	1.290	1.660	1.984	2.364	2.626	2.871	3.174
200	.676	1.286	1.652	1.972	2.345	2.601	2.838	3.131
500	.675	1.283	1.648	1.965	2.334	2.586	2.820	3.107
1,000	.675	1.282	1.646	1.962	2.330	2.581	2.813	3.098
2,000	.675	1.282	1.645	1.961	2.328	2.578	2.810	3.094
10,000	.675	1.282	1.645	1.960	2.327	2.576	2.808	3.091
∞	.674	1.282	1.645	1.960	2.326	2.576	2.807	3.090

SOURCE: Reprinted from Enrico T. Federighi, Extended tables of the percentage points of student's t distribution. *Journal of the American Statistical Association*, 1959, *54*, 683–688, by permission.

df \ P	.0005	.00025	.0001	.00005	.000025	.00001
1	636.619	1,273.239	3,183.099	6,366.198	12,732.395	31,830.989
2	31.598	44.705	70.700	99.992	141.416	223.603
3	12.924	16.326	22.204	28.000	35.298	47.928
4	8.610	10.306	13.034	15.544	18.522	23.332
5	6.869	7.976	9.678	11.178	12.893	15.547
6	5.959	6.788	8.025	9.082	10.261	12.032
7	5.408	6.082	7.063	7.885	8.782	10.103
8	5.041	5.618	6.442	7.120	7.851	8.907
9	4.781	5.291	6.010	6.594	7.215	8.102
10	4.587	5.049	5.694	6.211	6.757	7.527
11	4.437	4.863	5.453	5.921	6.412	7.098
12	4.318	4.716	5.263	5.694	6.143	6.756
13	4.221	4.597	5.111	5.513	5.928	6.501
14	4.140	4.499	4.985	5.363	5.753	6.287
15	4.073	4.417	4.880	5.239	5.607	6.109
16	4.015	4.346	4.791	5.134	5.484	5.960
17	3.965	4.286	4.714	5.044	5.379	5.832
18	3.922	4.233	4.648	4.966	5.288	5.722
19	3.883	4.187	4.590	4.897	5.209	5.627
20	3.850	4.146	4.539	4.837	5.139	5.543
21	3.819	4.110	4.493	4.784	5.077	5.469
22	3.792	4.077	4.452	4.736	5.022	5.402
23	3.768	4.048	4.415	4.693	4.972	5.343
24	3.745	4.021	4.382	4.654	4.927	5.290
25	3.725	3.997	4.352	4.619	4.887	5.241
26	3.707	3.974	4.324	4.587	4.850	5.197
27	3.690	3.954	4.299	4.558	4.816	5.157
28	3.674	3.935	4.275	4.530	4.784	5.120
29	3.659	3.918	4.254	4.506	4.756	5.086
30	3.646	3.902	4.234	4.482	4.729	5.054
35	3.591	3.836	4.153	4.389	4.622	4.927
40	3.551	3.788	4.094	4.321	4.544	4.835
45	3.520	3.752	4.049	4.269	4.485	4.766
50	3.496	3.723	4.014	4.228	4.438	4.711
55	3.476	3.700	3.986	4.196	4.401	4.667
60	3.460	3.681	3.962	4.169	4.370	4.631
70	3.435	3.651	3.926	4.127	4.323	4.576
80	3.416	3.629	3.899	4.096	4.288	4.535
90	3.402	3.612	3.878	4.072	4.261	4.503
100	3.390	3.598	3.862	4.053	4.240	4.478
200	3.340	3.539	3.789	3.970	4.146	4.369
500	3.310	3.504	3.747	3.922	4.091	4.306
1,000	3.300	3.492	3.733	3.906	4.073	4.285
2,000	3.295	3.486	3.726	3.898	4.064	4.275
10,000	3.292	3.482	3.720	3.892	4.058	4.267
∞	3.291	3.481	3.719	3.891	4.056	4.265

(Continued)

df \ P	.000005	.0000025	.000001	.0000005	.00000025	.0000001
1	63,661.977	127,323.954	318,309.886	636,619.772	1,273,239.545	3,183,098.862
2	316.225	447.212	707.106	999.999	1,414.213	2,236.068
3	60,397	76.104	103.299	130.155	163.989	222.572
4	27.771	33.047	41.578	49.459	58.829	73.986
5	17.897	20.591	24.771	28.477	32.734	39.340
6	13.555	15.260	17.830	20.047	22.532	26.286
7	11.215	12.437	14.241	15.764	17.447	19.932
8	9.782	10.731	12.110	13.257	14.504	16.320
9	8.827	9.605	10.720	11.637	12.623	14.041
10	8.150	8.812	9.752	10.516	11.328	12.492
11	7.648	8.227	9.043	9.702	10.397	11.381
12	7.261	7.780	8.504	9.085	9.695	10.551
13	6.955	7.427	8.082	8.604	9.149	9.909
14	6.706	7.142	7.743	8.218	8.713	9.400
15	6.502	6.907	7.465	7.903	8.358	8.986
16	6.330	6.711	7.233	7.642	8.064	8.645
17	6.184	6.545	7.037	7.421	7.817	8.358
18	6.059	6.402	6.869	7.232	7.605	8.115
19	5.949	6.278	6.723	7.069	7.423	7.905
20	5.854	6.170	6.597	6.927	7.265	7.723
21	5.769	6.074	6.485	6.802	7.126	7.564
22	5.694	5.989	6.386	6.692	7.003	7.423
23	5.627	5.913	6.297	6.593	6.893	7.298
24	5.566	5.845	6.218	6.504	6.795	7.185
25	5.511	5.783	6.146	6.424	6.706	7.085
26	5.461	5.726	6.081	6.352	6.626	6.993
27	5.415	5.675	6.021	6.286	6.553	6.910
28	5.373	5.628	5.967	6.225	6.486	6.835
39	5.335	5.585	5.917	6.170	6.426	6.765
30	5.299	5.545	5.871	6.119	6.369	6.701
35	5.156	5.385	5.687	5.915	6.143	6.447
40	5.053	5.269	5.554	5.768	5.983	6.266
45	4.975	5.182	5.454	5.659	5.862	6.130
50	4.914	5.115	5.377	5.573	5.769	6.025
55	4.865	5.060	5.315	5.505	5.694	5.942
60	4.825	5.015	5.264	5.449	5.633	5.873
70	4.763	4.946	5.185	5.363	5.539	5.768
80	4.717	4.896	5.128	5.300	5.470	5.691
90	4.682	4.857	5.084	5.252	5.417	5.633
100	4.654	4.826	5.049	5.214	5.376	5.587
200	4.533	4.692	4.897	5.048	5.196	5.387
500	4.463	4.615	4.810	4.953	5.094	5.273
1,000	4.440	4.590	4.781	4.922	5.060	5.236
2,000	4.428	4.578	4.767	4.907	5.043	5.218
10,000	4.419	4.567	4.756	4.895	5.029	5.203
∞	4.417	4.565	4.753	4.892	5.026	5.199

TABLE V Values of the Correlation Coefficient for Various Levels of Significance

The probabilities given by the column headings are for obtaining r equal to or greater than the tabled value, that is, for a one-sided test, when the null hypothesis $\rho = 0$ is true. For a two-sided test, the probabilities should be doubled.

df \ P	.050	.025	.010	.005
1	.988	.997	.9995	.9999
2	.900	.950	.980	.990
3	.805	.878	.934	.959
4	.729	.811	.882	.917
5	.669	.754	.833	.874
6	.622	.707	.789	.834
7	.582	.666	.750	.798
8	.549	.632	.716	.765
9	.521	.602	.685	.735
10	.497	.576	.658	.708
11	.476	.553	.634	.684
12	.458	.532	.612	.661
13	.441	.514	.592	.641
14	.426	.497	.574	.623
15	.412	.482	.558	.606
16	.400	.468	.542	.590
17	.389	.456	.528	.575
18	.378	.444	.516	.561
19	.369	.433	.503	.549
20	.360	.423	.492	.537
21	.352	.413	.482	.526
22	.344	.404	.472	.515
23	.337	.396	.462	.505
24	.330	.388	.453	.496
25	.323	.381	.445	.487
26	.317	.374	.437	.479
27	.311	.367	.430	.471
28	.306	.361	.423	.463
29	.301	.355	.416	.456
30	.296	.349	.409	.449
35	.275	.325	.381	.418
40	.257	.304	.358	.393
45	.243	.288	.338	.372
50	.231	.273	.322	.354
60	.211	.250	.295	.325
70	.195	.232	.274	.302
80	.183	.217	.256	.283
90	.173	.205	.242	.267
100	.164	.195	.230	.254

Additional values of r at the .025 and .005 levels of significance

df	.025	.005	df	.025	.005	df	.025	.005
32	.339	.436	48	.279	.361	150	.159	.208
34	.329	.424	55	.261	.338	175	.148	.193
36	.320	.413	65	.241	.313	200	.138	.181
38	.312	.403	75	.224	.292	300	.113	.148
42	.297	.384	85	.211	.275	400	.098	.128
44	.291	.376	95	.200	.260	500	.088	.115
46	.284	.368	125	.174	.228	1,000	.062	.081

SOURCE: Reprinted from Table VA of R. A. Fisher, *Statistical Methods for Research Workers* (14th ed.). Copyright © 1972 by Hafner Press, by permission of the publisher. Additional entries were calculated using the t distribution.

TABLE VI Table of values of $z_r = \frac{1}{2}[\log_e(1 + r) - \log_e(1 - r)]$

r	z_r	r	z_r	r	z_r	r	z_r	r	z_r
.000	.000	.200	.203	.400	.424	.600	.693	.800	1.099
.005	.005	.205	.208	.405	.430	.605	.701	.805	1.113
.010	.010	.210	.213	.410	.436	.610	.709	.810	1.127
.015	.015	.215	.218	.415	.442	.615	.717	.815	1.142
.020	.020	.220	.224	.420	.448	.620	.725	.820	1.157
.025	.025	.225	.229	.425	.454	.625	.733	.825	1.172
.030	.030	.230	.234	.430	.460	.630	.741	.830	1.188
.035	.035	.235	.239	.435	.466	.635	.750	.835	1.204
.040	.040	.240	.245	.440	.472	.640	.758	.840	1.221
.045	.045	.245	.250	.445	.478	.645	.767	.845	1.238
.050	.050	.250	.255	.450	.485	.650	.775	.850	1.256
.055	.055	.255	.261	.455	.491	.655	.784	.855	1.274
.060	.060	.260	.266	.460	.497	.660	.793	.860	1.293
.065	.065	.265	.271	.465	.504	.665	.802	.865	1.313
.070	.070	.270	.277	.470	.510	.670	.811	.870	1.333
.075	.075	.275	.282	.475	.517	.675	.820	.875	1.354
.080	.080	.280	.288	.480	.523	.680	.829	.880	1.376
.085	.085	.285	.293	.485	.530	.685	.838	.885	1.398
.090	.090	.290	.299	.490	.536	.690	.848	.890	1.422
.095	.095	.295	.304	.495	.543	.695	.858	.895	1.447
.100	.100	.300	.310	.500	.549	.700	.867	.900	1.472
.105	.105	.305	.315	.505	.556	.705	.877	.905	1.499
.110	.110	.310	.321	.510	.563	.710	.887	.910	1.528
.115	.116	.315	.326	.515	.570	.715	.897	.915	1.557
.120	.121	.320	.332	.520	.576	.720	.908	.920	1.589
.125	.126	.325	.337	.525	.583	.725	.918	.925	1.623
.130	.131	.330	.343	.530	.590	.730	.929	.930	1.658
.135	.136	.335	.348	.535	.597	.735	.940	.935	1.697
.140	.141	.340	.354	.540	.604	.740	.950	.940	1.738
.145	.146	.345	.360	.545	.611	.745	.962	.945	1.783
.150	.151	.350	.365	.550	.618	.750	.973	.950	1.832
.155	.156	.355	.371	.555	.626	.755	.984	.955	1.886
.160	.161	.360	.377	.560	.633	.760	.996	.960	1.946
.165	.167	.365	.383	.565	.640	.765	1.008	.965	2.014
.170	.172	.370	.388	.570	.648	.770	1.020	.970	2.092
.175	.177	.375	.394	.575	.655	.775	1.033	.975	2.185
.180	.182	.380	.400	.580	.662	.780	1.045	.980	2.298
.185	.187	.385	.406	.585	.670	.785	1.058	.985	2.443
.190	.192	.390	.412	.590	.678	.790	1.071	.990	2.647
.195	.198	.395	.418	.595	.685	.795	1.085	.995	2.994

SOURCE: Table VI was constructed by F. P. Kilpatrick and D. A. Buchanan.

TABLE VII Values of F significant with $\alpha = .05$ and $\alpha = .01$

The values of F significant at the .05 (roman type) and .01 (boldface type) levels of significance with n_1 degrees of freedom for the numerator and n_2 degrees of freedom for the denominator of the F ratio.

	n_1 Degrees of freedom																							
n_2	1	2	3	4	5	6	7	8	9	10	11	12	14	16	20	24	30	40	50	75	100	200	500	∞
1	161	200	216	225	230	234	237	239	241	242	243	244	245	246	248	249	250	251	252	253	253	254	254	254
	4,052	**4,999**	**5,403**	**5,625**	**5,764**	**5,859**	**5,928**	**5,981**	**6,022**	**6,056**	**6,082**	**6,106**	**6,142**	**6,169**	**6,208**	**6,234**	**6,258**	**6,286**	**6,302**	**6,323**	**6,334**	**6,352**	**6,361**	**6,366**
2	18.51	19.00	19.16	19.25	19.30	19.33	19.36	19.37	19.38	19.39	19.40	19.41	19.42	19.43	19.44	19.45	19.46	19.47	19.47	19.48	19.49	19.49	19.50	19.50
	98.49	**99.00**	**99.17**	**99.25**	**99.30**	**99.33**	**99.34**	**99.36**	**99.38**	**99.40**	**99.41**	**99.42**	**99.43**	**99.44**	**99.45**	**99.46**	**99.47**	**99.48**	**99.48**	**99.49**	**99.49**	**99.49**	**99.50**	**99.50**
3	10.13	9.55	9.28	9.12	9.01	8.94	8.88	8.84	8.81	8.78	8.76	8.74	8.71	8.69	8.66	8.64	8.62	8.60	8.58	8.57	8.56	8.54	8.54	8.53
	34.12	**30.82**	**29.46**	**28.71**	**28.24**	**27.91**	**27.67**	**27.49**	**27.34**	**27.23**	**27.13**	**27.05**	**26.92**	**26.83**	**26.69**	**26.60**	**26.50**	**26.41**	**26.35**	**26.27**	**26.23**	**26.18**	**26.14**	**26.12**
4	7.71	6.94	6.59	6.39	6.26	6.16	6.09	6.04	6.00	5.96	5.93	5.91	5.87	5.84	5.80	5.77	5.74	5.71	5.70	5.68	5.66	5.65	5.64	5.63
	21.20	**18.00**	**16.69**	**15.98**	**15.52**	**15.21**	**14.98**	**14.80**	**14.66**	**14.54**	**14.45**	**14.37**	**14.24**	**14.15**	**14.02**	**13.93**	**13.83**	**13.74**	**13.69**	**13.61**	**13.57**	**13.52**	**13.48**	**13.46**
5	6.61	5.79	5.41	5.19	5.05	4.95	4.88	4.82	4.78	4.74	4.70	4.68	4.64	4.60	4.56	4.53	4.50	4.46	4.44	4.42	4.40	4.38	4.37	4.36
	16.26	**13.27**	**12.06**	**11.39**	**10.97**	**10.67**	**10.45**	**10.27**	**10.15**	**10.05**	**9.96**	**9.89**	**9.77**	**9.68**	**9.55**	**9.47**	**9.38**	**9.29**	**9.24**	**9.17**	**9.13**	**9.07**	**9.04**	**9.02**
6	5.99	5.14	4.76	4.53	4.39	4.28	4.21	4.15	4.10	4.06	4.03	4.00	3.96	3.92	3.87	3.84	3.81	3.77	3.75	3.72	3.71	3.69	3.68	3.67
	13.74	**10.92**	**9.78**	**9.15**	**8.75**	**8.47**	**8.26**	**8.10**	**7.98**	**7.87**	**7.79**	**7.72**	**7.60**	**7.52**	**7.39**	**7.31**	**7.23**	**7.14**	**7.09**	**7.02**	**6.99**	**6.94**	**6.90**	**6.88**
7	5.59	4.74	4.35	4.12	3.97	3.87	3.79	3.73	3.68	3.63	3.60	3.57	3.52	3.49	3.44	3.41	3.38	3.34	3.32	3.29	3.28	3.25	3.24	3.23
	12.25	**9.55**	**8.45**	**7.85**	**7.46**	**7.19**	**7.00**	**6.84**	**6.71**	**6.62**	**6.54**	**6.47**	**6.35**	**6.27**	**6.15**	**6.07**	**5.98**	**5.90**	**5.85**	**5.78**	**5.75**	**5.70**	**5.67**	**5.65**
8	5.32	4.46	4.07	3.84	3.69	3.58	3.50	3.44	3.39	3.34	3.31	3.28	3.23	3.20	3.15	3.12	3.08	3.05	3.03	3.00	2.98	2.96	2.94	2.93
	11.26	**8.65**	**7.59**	**7.01**	**6.63**	**6.37**	**6.19**	**6.03**	**5.91**	**5.82**	**5.74**	**5.67**	**5.56**	**5.48**	**5.36**	**5.28**	**5.20**	**5.11**	**5.06**	**5.00**	**4.96**	**4.91**	**4.88**	**4.86**
9	5.12	4.26	3.86	3.63	3.48	3.37	3.29	3.23	3.18	3.13	3.10	3.07	3.02	2.98	2.93	2.90	2.86	2.82	2.80	2.77	2.76	2.73	2.72	2.71
	10.56	**8.02**	**6.99**	**6.42**	**6.06**	**5.80**	**5.62**	**5.47**	**5.35**	**5.26**	**5.18**	**5.11**	**5.00**	**4.92**	**4.80**	**4.73**	**4.64**	**4.56**	**4.51**	**4.45**	**4.41**	**4.36**	**4.33**	**4.31**
10	4.96	4.10	3.71	3.48	3.33	3.22	3.14	3.07	3.02	2.97	2.94	2.91	2.86	2.82	2.77	2.74	2.70	2.67	2.64	2.61	2.59	2.56	2.55	2.54
	10.04	**7.56**	**6.55**	**5.99**	**5.64**	**5.39**	**5.21**	**5.06**	**4.95**	**4.85**	**4.78**	**4.71**	**4.60**	**4.52**	**4.41**	**4.33**	**4.25**	**4.17**	**4.12**	**4.05**	**4.01**	**3.96**	**3.93**	**3.91**
11	4.84	3.98	3.59	3.36	3.20	3.09	3.01	2.95	2.90	2.86	2.82	2.79	2.74	2.70	2.65	2.61	2.57	2.53	2.50	2.47	2.45	2.42	2.41	2.40
	9.65	**7.20**	**6.22**	**5.67**	**5.32**	**5.07**	**4.88**	**4.74**	**4.63**	**4.54**	**4.46**	**4.40**	**4.29**	**4.21**	**4.10**	**4.02**	**3.94**	**3.86**	**3.80**	**3.74**	**3.70**	**3.66**	**3.62**	**3.60**
12	4.75	3.88	3.49	3.26	3.11	3.00	2.92	2.85	2.80	2.76	2.72	2.69	2.64	2.60	2.54	2.50	2.46	2.42	2.40	2.36	2.35	2.32	2.31	2.30
	9.33	**6.93**	**5.95**	**5.41**	**5.06**	**4.82**	**4.65**	**4.50**	**4.39**	**4.30**	**4.22**	**4.16**	**4.05**	**3.98**	**3.86**	**3.78**	**3.70**	**3.61**	**3.56**	**3.49**	**3.46**	**3.41**	**3.38**	**3.36**
13	4.67	3.80	3.41	3.18	3.02	2.92	2.84	2.77	2.72	2.67	2.63	2.60	2.55	2.51	2.46	2.42	2.38	2.34	2.32	2.28	2.26	2.24	2.22	2.21
	9.07	**6.70**	**5.74**	**5.20**	**4.86**	**4.62**	**4.44**	**4.30**	**4.19**	**4.10**	**4.02**	**3.96**	**3.85**	**3.78**	**3.67**	**3.59**	**3.51**	**3.42**	**3.37**	**3.30**	**3.27**	**3.21**	**3.18**	**3.16**
14	4.60	3.74	3.34	3.11	2.96	2.85	2.77	2.70	2.65	2.60	2.56	2.53	2.48	2.44	2.39	2.35	2.31	2.27	2.24	2.21	2.19	2.16	2.14	2.13
	8.86	**6.51**	**5.56**	**5.03**	**4.69**	**4.46**	**4.28**	**4.14**	**4.03**	**3.94**	**3.86**	**3.80**	**3.70**	**3.62**	**3.51**	**3.43**	**3.34**	**3.26**	**3.21**	**3.14**	**3.11**	**3.06**	**3.02**	**3.00**
15	4.54	3.68	3.29	3.06	2.90	2.79	2.70	2.64	2.59	2.55	2.51	2.48	2.43	2.39	2.33	2.29	2.25	2.21	2.18	2.15	2.12	2.10	2.08	2.07
	8.68	**6.36**	**5.42**	**4.89**	**4.56**	**4.32**	**4.14**	**4.00**	**3.89**	**3.80**	**3.73**	**3.67**	**3.56**	**3.48**	**3.36**	**3.29**	**3.20**	**3.12**	**3.07**	**3.00**	**2.97**	**2.92**	**2.89**	**2.87**
16	4.49	3.63	3.24	3.01	2.85	2.74	2.66	2.59	2.54	2.49	2.45	2.42	2.37	2.33	2.28	2.24	2.20	2.16	2.13	2.09	2.07	2.04	2.02	2.01
	8.53	**6.23**	**5.29**	**4.77**	**4.44**	**4.20**	**4.03**	**3.89**	**3.78**	**3.69**	**3.61**	**3.55**	**3.45**	**3.37**	**3.25**	**3.18**	**3.10**	**3.01**	**2.96**	**2.89**	**2.86**	**2.80**	**2.77**	**2.75**

SOURCE: Reprinted by permission from *Statistical Methods* (6th ed.), by George W. Snedecor and William G. Cochran.

(*Continued*)

	n_1 Degrees of freedom																							
n_2	1	2	3	4	5	6	7	8	9	10	11	12	14	16	20	24	30	40	50	75	100	200	500	∞
17	4.45	3.59	3.20	2.96	2.81	2.70	2.62	2.55	2.50	2.45	2.41	2.38	2.33	2.29	2.23	2.19	2.15	2.11	2.08	2.04	2.02	1.99	1.97	1.96
	8.40	**6.11**	**5.18**	**4.67**	**4.34**	**4.10**	**3.93**	**3.79**	**3.68**	**3.59**	**3.52**	**3.45**	**3.35**	**3.27**	**3.16**	**3.08**	**3.00**	**2.92**	**2.86**	**2.79**	**2.76**	**2.70**	**2.67**	**2.65**
18	4.41	3.55	3.16	2.93	2.77	2.66	2.58	2.51	2.46	2.41	2.37	2.34	2.29	2.25	2.19	2.15	2.11	2.07	2.04	2.00	1.98	1.95	1.93	1.92
	8.28	**6.01**	**5.09**	**4.58**	**4.25**	**4.01**	**3.85**	**3.71**	**3.60**	**3.51**	**3.44**	**3.37**	**3.27**	**3.19**	**3.07**	**3.00**	**2.91**	**2.83**	**2.78**	**2.71**	**2.68**	**2.62**	**2.59**	**2.57**
19	4.38	3.52	3.13	2.90	2.74	2.63	2.55	2.48	2.43	2.38	2.34	2.31	2.26	2.21	2.15	2.11	2.07	2.02	2.00	1.96	1.94	1.91	1.90	1.88
	8.18	**5.93**	**5.01**	**4.50**	**4.17**	**3.94**	**3.77**	**3.63**	**3.52**	**3.43**	**3.36**	**3.30**	**3.19**	**3.12**	**3.00**	**2.92**	**2.84**	**2.76**	**2.70**	**2.63**	**2.60**	**2.54**	**2.51**	**2.49**
20	4.35	3.49	3.10	2.87	2.71	2.60	2.52	2.45	2.40	2.35	2.31	2.28	2.23	2.18	2.12	2.08	2.04	1.99	1.96	1.92	1.90	1.87	1.85	1.84
	8.10	**5.85**	**4.94**	**4.43**	**4.10**	**3.87**	**3.71**	**3.56**	**3.45**	**3.37**	**3.30**	**3.23**	**3.13**	**3.05**	**2.94**	**2.86**	**2.77**	**2.69**	**2.63**	**2.56**	**2.53**	**2.47**	**2.44**	**2.42**
21	4.32	3.47	3.07	2.84	2.68	2.57	2.49	2.42	2.37	2.32	2.28	2.25	2.20	2.15	2.09	2.05	2.00	1.96	1.93	1.89	1.87	1.84	1.82	1.81
	8.02	**5.78**	**4.87**	**4.37**	**4.04**	**3.81**	**3.65**	**3.51**	**3.40**	**3.31**	**3.24**	**3.17**	**3.07**	**2.99**	**2.88**	**2.80**	**2.72**	**2.63**	**2.58**	**2.51**	**2.47**	**2.42**	**2.38**	**2.36**
22	4.30	3.44	3.05	2.82	2.66	2.55	2.47	2.40	2.35	2.30	2.26	2.23	2.18	2.13	2.07	2.03	1.98	1.93	1.91	1.87	1.84	1.81	1.80	1.78
	7.94	**5.72**	**4.82**	**4.31**	**3.99**	**3.76**	**3.59**	**3.45**	**3.35**	**3.26**	**3.18**	**3.12**	**3.02**	**2.94**	**2.83**	**2.75**	**2.67**	**2.58**	**2.53**	**2.46**	**2.42**	**2.37**	**2.33**	**2.31**
23	4.28	3.42	3.03	2.80	2.64	2.53	2.45	2.38	2.32	2.28	2.24	2.20	2.14	2.10	2.04	2.00	1.96	1.91	1.88	1.84	1.82	1.79	1.77	1.76
	7.88	**5.66**	**4.76**	**4.26**	**3.94**	**3.71**	**3.54**	**3.41**	**3.30**	**3.21**	**3.14**	**3.07**	**2.97**	**2.89**	**2.78**	**2.70**	**2.62**	**2.53**	**2.48**	**2.41**	**2.37**	**2.32**	**2.28**	**2.26**
24	4.26	3.40	3.01	2.78	2.62	2.51	2.43	2.36	2.30	2.26	2.22	2.18	2.13	2.09	2.02	1.98	1.94	1.89	1.86	1.82	1.80	1.76	1.74	1.73
	7.82	**5.61**	**4.72**	**4.22**	**3.90**	**3.67**	**3.50**	**3.36**	**3.25**	**3.17**	**3.09**	**3.03**	**2.93**	**2.85**	**2.74**	**2.66**	**2.58**	**2.49**	**2.44**	**2.36**	**2.33**	**2.27**	**2.23**	**2.21**
25	4.24	3.38	2.99	2.76	2.60	2.49	2.41	2.34	2.28	2.24	2.20	2.16	2.11	2.06	2.00	1.96	1.92	1.87	1.84	1.80	1.77	1.74	1.72	1.71
	7.77	**5.57**	**4.68**	**4.18**	**3.86**	**3.63**	**3.46**	**3.32**	**3.21**	**3.13**	**3.05**	**2.99**	**2.89**	**2.81**	**2.70**	**2.62**	**2.54**	**2.45**	**2.40**	**2.32**	**2.29**	**2.23**	**2.19**	**2.17**
26	4.22	3.37	2.98	2.74	2.59	2.47	2.39	2.32	2.27	2.22	2.18	2.15	2.10	2.05	1.99	1.95	1.90	1.85	1.82	1.78	1.76	1.72	1.70	1.69
	7.72	**5.53**	**4.64**	**4.14**	**3.82**	**3.59**	**3.42**	**3.29**	**3.17**	**3.09**	**3.02**	**2.96**	**2.86**	**2.77**	**2.66**	**2.58**	**2.50**	**2.41**	**2.36**	**2.28**	**2.25**	**2.19**	**2.15**	**2.13**
27	4.21	3.35	2.96	2.73	2.57	2.46	2.37	2.30	2.25	2.20	2.16	2.13	2.08	2.03	1.97	1.93	1.88	1.84	1.80	1.76	1.74	1.71	1.68	1.67
	7.68	**5.49**	**4.60**	**4.11**	**3.79**	**3.56**	**3.39**	**3.26**	**3.14**	**3.06**	**2.98**	**2.93**	**2.83**	**2.74**	**2.63**	**2.55**	**2.47**	**2.38**	**2.33**	**2.25**	**2.21**	**2.16**	**2.12**	**2.10**
28	4.20	3.34	2.95	2.71	2.56	2.44	2.36	2.29	2.24	2.19	2.15	2.12	2.06	2.02	1.96	1.91	1.87	1.81	1.78	1.75	1.72	1.69	1.67	1.65
	7.64	**5.45**	**4.57**	**4.07**	**3.76**	**3.53**	**3.36**	**3.23**	**3.11**	**3.03**	**2.95**	**2.90**	**2.80**	**2.71**	**2.60**	**2.52**	**2.44**	**2.35**	**2.30**	**2.22**	**2.18**	**2.13**	**2.09**	**2.06**
29	4.18	3.33	2.93	2.70	2.54	2.43	2.35	2.28	2.22	2.18	2.14	2.10	2.05	2.00	1.94	1.90	1.85	1.80	1.77	1.73	1.71	1.68	1.65	1.64
	7.60	**5.42**	**4.54**	**4.04**	**3.73**	**3.50**	**3.33**	**3.20**	**3.08**	**3.00**	**2.92**	**2.87**	**2.77**	**2.68**	**2.57**	**2.49**	**2.41**	**2.32**	**2.27**	**2.19**	**2.15**	**2.10**	**2.06**	**2.03**
30	4.17	3.32	2.92	2.69	2.53	2.42	2.34	2.27	2.21	2.16	2.12	2.09	2.04	1.99	1.93	1.89	1.84	1.79	1.76	1.72	1.69	1.66	1.64	1.62
	7.56	**5.39**	**4.51**	**4.02**	**3.70**	**3.47**	**3.30**	**3.17**	**3.06**	**2.98**	**2.90**	**2.84**	**2.74**	**2.66**	**2.55**	**2.47**	**2.38**	**2.29**	**2.24**	**2.16**	**2.13**	**2.07**	**2.03**	**2.01**
32	4.15	3.30	2.90	2.67	2.51	2.40	2.32	2.25	2.19	2.14	2.10	2.07	2.02	1.97	1.91	1.86	1.82	1.76	1.74	1.69	1.67	1.64	1.61	1.59
	7.50	**5.34**	**4.46**	**3.97**	**3.66**	**3.42**	**3.25**	**3.12**	**3.01**	**2.94**	**2.86**	**2.80**	**2.70**	**2.62**	**2.51**	**2.42**	**2.34**	**2.25**	**2.20**	**2.12**	**2.08**	**2.02**	**1.98**	**1.96**
34	4.13	3.28	2.88	2.65	2.49	2.38	2.30	2.23	2.17	2.12	2.08	2.05	2.00	1.95	1.89	1.84	1.80	1.74	1.71	1.67	1.64	1.61	1.59	1.57
	7.44	**5.29**	**4.42**	**3.93**	**3.61**	**3.38**	**3.21**	**3.08**	**2.97**	**2.89**	**2.82**	**2.76**	**2.66**	**2.58**	**2.47**	**2.38**	**2.30**	**2.21**	**2.15**	**2.08**	**2.04**	**1.98**	**1.94**	**1.91**
36	4.11	3.26	2.86	2.63	2.48	2.36	2.28	2.21	2.15	2.10	2.06	2.03	1.98	1.93	1.87	1.82	1.78	1.72	1.69	1.65	1.62	1.59	1.56	1.55
	7.39	**5.25**	**4.38**	**3.89**	**3.58**	**3.35**	**3.18**	**3.04**	**2.94**	**2.86**	**2.78**	**2.72**	**2.62**	**2.54**	**2.43**	**2.35**	**2.26**	**2.17**	**2.12**	**2.04**	**2.00**	**1.94**	**1.90**	**1.87**
38	4.10	3.25	2.85	2.62	2.46	2.35	2.26	2.19	2.14	2.09	2.05	2.02	1.96	1.92	1.85	1.80	1.76	1.71	1.67	1.63	1.60	1.57	1.54	1.53
	7.35	**5.21**	**4.34**	**3.86**	**3.54**	**3.32**	**3.15**	**3.02**	**2.91**	**2.82**	**2.75**	**2.69**	**2.59**	**2.51**	**2.40**	**2.32**	**2.22**	**2.14**	**2.08**	**2.00**	**1.97**	**1.90**	**1.86**	**1.84**

n_2	n_1 Degrees of freedom																							
	1	2	3	4	5	6	7	8	9	10	11	12	14	16	20	24	30	40	50	75	100	200	500	∞
40	4.08 **7.31**	3.23 **5.18**	2.84 **4.31**	2.61 **3.83**	2.45 **3.51**	2.34 **3.29**	2.25 **3.12**	2.18 **2.99**	2.12 **2.88**	2.07 **2.80**	2.04 **2.73**	2.00 **2.66**	1.95 **2.56**	1.90 **2.49**	1.84 **2.37**	1.79 **2.29**	1.74 **2.20**	1.69 **2.11**	1.66 **2.05**	1.61 **1.97**	1.59 **1.94**	1.55 **1.88**	1.53 **1.84**	1.51 **1.81**
42	4.07 **7.27**	3.22 **5.15**	2.83 **4.29**	2.59 **3.80**	2.44 **3.49**	2.32 **3.26**	2.24 **3.10**	2.17 **2.96**	2.11 **2.86**	2.06 **2.77**	2.02 **2.70**	1.99 **2.64**	1.94 **2.54**	1.89 **2.46**	1.82 **2.35**	1.78 **2.26**	1.73 **2.17**	1.68 **2.08**	1.64 **2.02**	1.60 **1.94**	1.57 **1.91**	1.54 **1.85**	1.51 **1.80**	1.49 **1.78**
44	4.06 **7.24**	3.21 **5.12**	2.82 **4.26**	2.58 **3.78**	2.43 **3.46**	2.31 **3.24**	2.23 **3.07**	2.16 **2.94**	2.10 **2.84**	2.05 **2.75**	2.01 **2.68**	1.98 **2.62**	1.92 **2.52**	1.88 **2.44**	1.81 **2.32**	1.76 **2.24**	1.72 **2.15**	1.66 **2.06**	1.63 **2.00**	1.58 **1.92**	1.56 **1.88**	1.52 **1.82**	1.50 **1.78**	1.48 **1.75**
46	4.05 **7.21**	3.20 **5.10**	2.81 **4.24**	2.57 **3.76**	2.42 **3.44**	2.30 **3.22**	2.22 **3.05**	2.14 **2.92**	2.09 **2.82**	2.04 **2.73**	2.00 **2.66**	1.97 **2.60**	1.91 **2.50**	1.87 **2.42**	1.80 **2.30**	1.75 **2.22**	1.71 **2.13**	1.65 **2.04**	1.62 **1.98**	1.57 **1.90**	1.54 **1.86**	1.51 **1.80**	1.48 **1.76**	1.46 **1.72**
48	4.04 **7.19**	3.19 **5.08**	2.80 **4.22**	2.56 **3.74**	2.41 **3.42**	2.30 **3.20**	2.21 **3.04**	2.14 **2.90**	2.08 **2.80**	2.03 **2.71**	1.99 **2.64**	1.96 **2.58**	1.90 **2.48**	1.86 **2.40**	1.79 **2.28**	1.74 **2.20**	1.70 **2.11**	1.64 **2.02**	1.61 **1.96**	1.56 **1.88**	1.53 **1.84**	1.50 **1.78**	1.47 **1.73**	1.45 **1.70**
50	4.03 **7.17**	3.18 **5.06**	2.79 **4.20**	2.56 **3.72**	2.40 **3.41**	2.29 **3.18**	2.20 **3.02**	2.13 **2.88**	2.07 **2.78**	2.02 **2.70**	1.98 **2.62**	1.95 **2.56**	1.90 **2.46**	1.85 **2.39**	1.78 **2.26**	1.74 **2.18**	1.69 **2.10**	1.63 **2.00**	1.60 **1.94**	1.55 **1.86**	1.52 **1.82**	1.48 **1.76**	1.46 **1.71**	1.44 **1.68**
55	4.02 **7.12**	3.17 **5.01**	2.78 **4.16**	2.54 **3.68**	2.38 **3.37**	2.27 **3.15**	2.18 **2.98**	2.11 **2.85**	2.05 **2.75**	2.00 **2.66**	1.97 **2.59**	1.93 **2.53**	1.88 **2.43**	1.83 **2.35**	1.76 **2.23**	1.72 **2.15**	1.67 **2.06**	1.61 **1.96**	1.58 **1.90**	1.52 **1.82**	1.50 **1.78**	1.46 **1.71**	1.43 **1.66**	1.41 **1.64**
60	4.00 **7.08**	3.15 **4.98**	2.76 **4.13**	2.52 **3.65**	2.37 **3.34**	2.25 **3.12**	2.17 **2.95**	2.10 **2.82**	2.04 **2.72**	1.99 **2.63**	1.95 **2.56**	1.92 **2.50**	1.86 **2.40**	1.81 **2.32**	1.75 **2.20**	1.70 **2.12**	1.65 **2.03**	1.59 **1.93**	1.56 **1.87**	1.50 **1.79**	1.48 **1.74**	1.44 **1.68**	1.41 **1.63**	1.39 **1.60**
65	3.99 **7.04**	3.14 **4.95**	2.75 **4.10**	2.51 **3.62**	2.36 **3.31**	2.24 **3.09**	2.15 **2.93**	2.08 **2.79**	2.02 **2.70**	1.98 **2.61**	1.94 **2.54**	1.90 **2.47**	1.85 **2.37**	1.80 **2.30**	1.73 **2.18**	1.68 **2.09**	1.63 **2.00**	1.57 **1.90**	1.54 **1.84**	1.49 **1.76**	1.46 **1.71**	1.42 **1.64**	1.39 **1.60**	1.37 **1.56**
70	3.98 **7.01**	3.13 **4.92**	2.74 **4.08**	2.50 **3.60**	2.35 **3.29**	2.23 **3.07**	2.14 **2.91**	2.07 **2.77**	2.01 **2.67**	1.97 **2.59**	1.93 **2.51**	1.89 **2.45**	1.84 **2.35**	1.79 **2.28**	1.72 **2.15**	1.67 **2.07**	1.62 **1.98**	1.56 **1.88**	1.53 **1.82**	1.47 **1.74**	1.45 **1.69**	1.40 **1.62**	1.37 **1.56**	1.35 **1.53**
80	3.96 **6.96**	3.11 **4.88**	2.72 **4.04**	2.48 **3.56**	2.33 **3.25**	2.21 **3.04**	2.12 **2.87**	2.05 **2.74**	1.99 **2.64**	1.95 **2.55**	1.91 **2.48**	1.88 **2.41**	1.82 **2.32**	1.77 **2.24**	1.70 **2.11**	1.65 **2.03**	1.60 **1.94**	1.54 **1.84**	1.51 **1.78**	1.45 **1.70**	1.42 **1.65**	1.38 **1.57**	1.35 **1.52**	1.32 **1.49**
100	3.94 **6.90**	3.09 **4.82**	2.70 **3.98**	2.46 **3.51**	2.30 **3.20**	2.19 **2.99**	2.10 **2.82**	2.03 **2.69**	1.97 **2.59**	1.92 **2.51**	1.88 **2.43**	1.85 **2.36**	1.79 **2.26**	1.75 **2.19**	1.68 **2.06**	1.63 **1.98**	1.57 **1.89**	1.51 **1.79**	1.48 **1.73**	1.42 **1.64**	1.39 **1.59**	1.34 **1.51**	1.30 **1.46**	1.28 **1.43**
125	3.92 **6.84**	3.07 **4.78**	2.68 **3.94**	2.44 **3.47**	2.29 **3.17**	2.17 **2.95**	2.08 **2.79**	2.01 **2.65**	1.95 **2.56**	1.90 **2.47**	1.86 **2.40**	1.83 **2.33**	1.77 **2.23**	1.72 **2.15**	1.65 **2.03**	1.60 **1.94**	1.55 **1.85**	1.49 **1.75**	1.45 **1.68**	1.39 **1.59**	1.36 **1.54**	1.31 **1.46**	1.27 **1.40**	1.25 **1.37**
150	3.91 **6.81**	3.06 **4.75**	2.67 **3.91**	2.43 **3.44**	2.27 **3.14**	2.16 **2.92**	2.07 **2.76**	2.00 **2.62**	1.94 **2.53**	1.89 **2.44**	1.85 **2.37**	1.82 **2.30**	1.76 **2.20**	1.71 **2.12**	1.64 **2.00**	1.59 **1.91**	1.54 **1.83**	1.47 **1.72**	1.44 **1.66**	1.37 **1.56**	1.34 **1.51**	1.29 **1.43**	1.25 **1.37**	1.22 **1.33**
200	3.89 **6.76**	3.04 **4.71**	2.65 **3.88**	2.41 **3.41**	2.26 **3.11**	2.14 **2.90**	2.05 **2.73**	1.98 **2.60**	1.92 **2.50**	1.87 **2.41**	1.83 **2.34**	1.80 **2.28**	1.74 **2.17**	1.69 **2.09**	1.62 **1.97**	1.57 **1.88**	1.52 **1.79**	1.45 **1.69**	1.42 **1.62**	1.35 **1.53**	1.32 **1.48**	1.26 **1.39**	1.22 **1.33**	1.19 **1.28**
400	3.86 **6.70**	3.02 **4.66**	2.62 **3.83**	2.39 **3.36**	2.23 **3.06**	2.12 **2.85**	2.03 **2.69**	1.96 **2.55**	1.90 **2.46**	1.85 **2.37**	1.81 **2.29**	1.78 **2.23**	1.72 **2.12**	1.67 **2.04**	1.60 **1.92**	1.54 **1.84**	1.49 **1.74**	1.42 **1.64**	1.38 **1.57**	1.32 **1.47**	1.28 **1.42**	1.22 **1.32**	1.16 **1.24**	1.13 **1.19**
1000	3.85 **6.66**	3.00 **4.62**	2.61 **3.80**	2.38 **3.34**	2.22 **3.04**	2.10 **2.82**	2.02 **2.66**	1.95 **2.53**	1.89 **2.43**	1.84 **2.34**	1.80 **2.26**	1.76 **2.20**	1.70 **2.09**	1.65 **2.01**	1.58 **1.89**	1.53 **1.81**	1.47 **1.71**	1.41 **1.61**	1.36 **1.54**	1.30 **1.44**	1.26 **1.38**	1.19 **1.28**	1.13 **1.19**	1.08 **1.11**
∞	3.84 **6.64**	2.99 **4.60**	2.60 **3.78**	2.37 **3.32**	2.21 **3.02**	2.09 **2.80**	2.01 **2.64**	1.94 **2.51**	1.88 **2.41**	1.83 **2.32**	1.79 **2.24**	1.75 **2.18**	1.69 **2.07**	1.64 **1.99**	1.57 **1.87**	1.52 **1.79**	1.46 **1.69**	1.40 **1.59**	1.35 **1.52**	1.28 **1.41**	1.24 **1.36**	1.17 **1.25**	1.11 **1.15**	1.00 **1.00**

TABLE VIII Table of four-place logarithms

To obtain the mantissa for a four-digit number, find in the body of the table the mantissa for the first three digits and then, neglecting the decimal point temporarily, add the number in the proportionality-parts table at the right that is on the same line as the mantissa already obtained and in the column corresponding to the fourth digit.

N	0	1	2	3	4	5	6	7	8	9	1	2	3	4	5	6	7	8	9
1.0	.0000	.0043	.0086	.0128	.0170	.0212	.0253	.0294	.0334	.0374	4	8	12	17	21	25	29	33	37
1.1	.0414	.0453	.0492	.0531	.0569	.0607	.0645	.0682	.0719	.0755	4	8	11	15	19	23	26	30	34
1.2	.0792	.0828	.0864	.0899	.0934	.0969	.1004	.1038	.1072	.1106	3	7	10	14	17	21	24	28	31
1.3	.1139	.1173	.1206	.1239	.1271	.1303	.1335	.1367	.1399	.1430	3	6	10	13	16	19	23	26	29
1.4	.1461	.1492	.1523	.1553	.1584	.1614	.1644	.1673	.1703	.1732	3	6	9	12	15	18	21	24	27
1.5	.1761	.1790	.1818	.1847	.1875	.1903	.1931	.1959	.1987	.2014	3	6	8	11	14	17	20	22	25
1.6	.2041	.2068	.2095	.2122	.2148	.2175	.2201	.2227	.2253	.2279	3	5	8	11	13	16	18	21	24
1.7	.2304	.2330	.2355	.2380	.2405	.2430	.2455	.2480	.2504	.2529	2	5	7	10	12	15	17	20	22
1.8	.2553	.2577	.2601	.2625	.2648	.2672	.2695	.2718	.2742	.2765	2	5	7	9	12	14	16	19	21
1.9	.2788	.2810	.2833	.2856	.2878	.2900	.2923	.2945	.2967	.2989	2	4	7	9	11	13	16	18	20
2.0	.3010	.3032	.3054	.3075	.3096	.3118	.3139	.3160	.3181	.3201	2	4	6	8	11	13	15	17	19
2.1	.3222	.3243	.3263	.3284	.3304	.3324	.3345	.3365	.3385	.3404	2	4	6	8	10	12	14	16	18
2.2	.3424	.3444	.3464	.3483	.3502	.3522	.3541	.3560	.3579	.3598	2	4	6	8	10	12	14	15	17
2.3	.3617	.3636	.3655	.3674	.3692	.3711	.3729	.3747	.3766	.3784	2	4	6	7	9	11	13	15	17
2.4	.3802	.3820	.3838	.3856	.3874	.3892	.3909	.3927	.3945	.3962	2	4	5	7	9	11	12	14	16
2.5	.3979	.3997	.4014	.4031	.4048	.4065	.4082	.4099	.4116	.4133	2	3	5	7	9	10	12	14	15
2.6	.4150	.4166	.4183	.4200	.4216	.4232	.4249	.4265	.4281	.4298	2	3	5	7	8	10	11	13	15
2.7	.4314	.4330	.4346	.4362	.4378	.4393	.4409	.4425	.4440	.4456	2	3	5	6	8	9	11	13	14
2.8	.4472	.4487	.4502	.4518	.4533	.4548	.4564	.4579	.4594	.4609	2	3	5	6	8	9	11	12	14
2.9	.4624	.4639	.4654	.4669	.4683	.4698	.4713	.4728	.4742	.4757	1	3	4	6	7	9	10	12	13
3.0	.4771	.4786	.4800	.4814	.4829	.4843	.4857	.4871	.4886	.4900	1	3	4	6	7	9	10	11	13
3.1	.4914	.4928	.4942	.4955	.4969	.4983	.4997	.5011	.5024	.5038	1	3	4	6	7	8	10	11	12
3.2	.5051	.5065	.5079	.5092	.5105	.5119	.5132	.5145	.5159	.5172	1	3	4	5	7	8	9	11	12
3.3	.5185	.5198	.5211	.5224	.5237	.5250	.5263	.5276	.5289	.5302	1	3	4	5	6	8	9	10	12
3.4	.5315	.5328	.5340	.5353	.5366	.5378	.5391	.5403	.5416	.5428	1	3	4	5	6	8	9	10	11
3.5	.5441	.5453	.5465	.5478	.5490	.5502	.5514	.5527	.5539	.5551	1	2	4	5	6	7	9	10	11
3.6	.5563	.5575	.5587	.5599	.5611	.5623	.5635	.5647	.5658	.5670	1	2	4	5	6	7	8	10	11
3.7	.5682	.5694	.5705	.5717	.5729	.5740	.5752	.5763	.5775	.5786	1	2	3	5	6	7	8	9	10
3.8	.5798	.5809	.5821	.5832	.5843	.5855	.5866	.5877	.5888	.5899	1	2	3	5	6	7	8	9	10
3.9	.5911	.5922	.5933	.5944	.5955	.5966	.5977	.5988	.5999	.6010	1	2	3	4	5	7	8	9	10

4.0	.6021	.6031	.6042	.6053	.6064	.6075	.6085	.6096	.6107	.6117	1	2	3	4	5	6	8	9	10
4.1	.6128	.6138	.6149	.6160	.6170	.6180	.6191	.6201	.6212	.6222	1	2	3	4	5	6	7	8	9
4.2	.6232	.6243	.6253	.6263	.6274	.6284	.6294	.6304	.6314	.6325	1	2	3	4	5	6	7	8	9
4.3	.6335	.6345	.6355	.6365	.6375	.6385	.6395	.6405	.6415	.6425	1	2	3	4	5	6	7	8	9
4.4	.6435	.6444	.6454	.6464	.6474	.6484	.6493	.6503	.6513	.6522	1	2	3	4	5	6	7	8	9
4.5	.6532	.6542	.6551	.6561	.6571	.6580	.6590	.6599	.6609	.6618	1	2	3	4	5	6	7	8	9
4.6	.6628	.6637	.6646	.6656	.6665	.6675	.6684	.6693	.6702	.6712	1	2	3	4	5	6	7	7	8
4.7	.6721	.6730	.6739	.6749	.6758	.6767	.6776	.6785	.6794	.6803	1	2	3	4	5	5	6	7	8
4.8	.6812	.6821	.6830	.6839	.6848	.6857	.6866	.6875	.6884	.6893	1	2	3	4	4	5	6	7	8
4.9	.6902	.6911	.6920	.6928	.6937	.6946	.6955	.6964	.6972	.6981	1	2	3	4	4	5	6	7	8
5.0	.6990	.6998	.7007	.7016	.7024	.7033	.7042	.7050	.7059	.7067	1	2	3	3	4	5	6	7	8
5.1	.7076	.7084	.7093	.7101	.7110	.7118	.7126	.7135	.7143	.7152	1	2	3	3	4	5	6	7	8
5.2	.7160	.7168	.7177	.7185	.7193	.7202	.7210	.7218	.7226	.7235	1	2	2	3	4	5	6	7	7
5.3	.7243	.7251	.7259	.7267	.7275	.7284	.7292	.7300	.7308	.7316	1	2	2	3	4	5	6	6	7
5.4	.7324	.7332	.7340	.7348	.7356	.7364	.7372	.7380	.7388	.7396	1	2	2	3	4	5	6	6	7
5.5	.7404	.7412	.7419	.7427	.7435	.7443	.7451	.7459	.7466	.7474	1	2	2	3	4	5	5	6	7
5.6	.7482	.7490	.7497	.7505	.7513	.7520	.7528	.7536	.7543	.7551	1	2	2	3	4	5	5	6	7
5.7	.7559	.7566	.7674	.7582	.7589	.7597	.7604	.7612	.7619	.7627	1	2	2	3	4	5	5	6	7
5.8	.7634	.7642	.7649	.7657	.7664	.7672	.7679	.7686	.7694	.7701	1	1	2	3	4	4	5	6	7
5.9	.7709	.7716	.7723	.7731	.7738	.7745	.7752	.7760	.7767	.7774	1	1	2	3	4	4	5	6	7
6.0	.7782	.7789	.7796	.7802	.7810	.7818	.7825	.7832	.7839	.7846	1	1	2	3	4	4	5	6	6
6.1	.7853	.7860	.7868	.7875	.7882	.7889	.7896	.7903	.7910	.7917	1	1	2	3	4	4	5	6	6
6.2	.7924	.7931	.7938	.7945	.7952	.7959	.7966	.7973	.7980	.7987	1	1	2	3	3	4	5	6	6
6.3	.7993	.8000	.8007	.8014	.8021	.8028	.8035	.8041	.8048	.8055	1	1	2	3	3	4	5	5	6
6.4	.8062	.8069	.8075	.8082	.8089	.8096	.8102	.8109	.8116	.8122	1	1	2	3	3	4	5	5	6
6.5	.8129	.8136	.8142	.8149	.8156	.8162	.8169	.8176	.8182	.8189	1	1	2	3	3	4	5	5	6
6.6	.8195	.8202	.8209	.8215	.8222	.8228	.8235	.8241	.8248	.8254	1	1	2	3	3	4	5	5	6
6.7	.8261	.8267	.8274	.8280	.8287	.8293	.8299	.8306	.8312	.8319	1	1	2	3	3	4	5	5	6
6.8	.8325	.8331	.8338	.8344	.8351	.8357	.8363	.8370	.8376	.8382	1	1	2	3	3	4	4	5	6
6.9	.8388	.8395	.8401	.8407	.8414	.8420	.8426	.8432	.8439	.8445	1	1	2	2	3	4	4	5	6
7.0	.8451	.8457	.8463	.8470	.8476	.8482	.8488	.8494	.8500	.8506	1	1	2	2	3	4	4	5	6
7.1	.8513	.8519	.8525	.8531	.8537	.8543	.8549	.8555	.8651	.8567	1	1	2	2	3	4	4	5	5
7.2	.8573	.8579	.8585	.8591	.8597	.8603	.8609	.8615	.8621	.8627	1	1	2	2	3	4	4	5	5
7.3	.8633	.8639	.8645	.8651	.8657	.8663	.8669	.8675	.8681	.8686	1	1	2	2	3	4	4	5	5
7.4	.8692	.8698	.8704	.8710	.8716	.8722	.8727	.8733	.8739	.8745	1	1	2	2	3	4	4	5	5

(*Continued*)

N	0	1	2	3	4	5	6	7	8	9	1	2	3	4	5	6	7	8	9
7.5	.8751	.8756	.8762	.8768	.8774	.8779	.8785	.8791	.8797	.8802	1	1	2	2	3	3	4	5	5
7.6	.8808	.8814	.8820	.8925	.8831	.8837	.8842	.8848	.8854	.8859	1	1	2	2	3	3	4	5	5
7.7	.8865	.8871	.8876	.8882	.8887	.8893	.8899	.8904	.8910	.8915	1	1	2	2	3	3	4	4	5
7.8	.8921	.8927	.8932	.8938	.8943	.8949	.8954	.8960	.8965	.8971	1	1	2	2	3	3	4	4	5
7.9	.8976	.8982	.8987	.8993	.8998	.9004	.9009	.9015	.9020	.9025	1	1	2	2	3	3	4	4	5
8.0	.9031	.9036	.9042	.9047	.9053	.9058	.9063	.9069	.9074	.9079	1	1	2	2	3	3	4	4	5
8.1	.9085	.9090	.9096	.9101	.9106	.9112	.9117	.9122	.9128	.9133	1	1	2	2	3	3	4	4	5
8.2	.9138	.9143	.9149	.9154	.9159	.9165	.9170	.9175	.9180	.9186	1	1	2	2	3	3	4	4	5
8.3	.9191	.9196	.9201	.9206	.9212	.9217	.9222	.9227	.9232	.9238	1	1	2	2	3	3	4	4	5
8.4	.9243	.9248	.9253	.9258	.9263	.9269	.9274	.9279	.9284	.9289	1	1	2	2	3	3	4	4	5
8.5	.9294	.9299	.9304	.9309	.9315	.9320	.9325	.9330	.9335	.9340	1	1	2	2	3	3	4	4	5
8.6	.9345	.9350	.9355	.9360	.9365	.9370	.9375	.9380	.9385	.9390	1	1	2	2	3	3	4	4	5
8.7	.9395	.9400	.9405	.9410	.9415	.9420	.9425	.9430	.9435	.9440	0	1	1	2	2	3	3	4	4
8.8	.9445	.9450	.9455	.9460	.9465	.9469	.9474	.9479	.9484	.9489	0	1	1	2	2	3	3	4	4
8.9	.9494	.9499	.9504	.9509	.9513	.9518	.9523	.9528	.9533	.9538	0	1	1	2	2	3	3	4	4
9.0	.9542	.9547	.9552	.9557	.9562	.9566	.9571	.9576	.9581	.9586	0	1	1	2	2	3	3	4	4
9.1	.9590	.9595	.9600	.9605	.9609	.9614	.9619	.9624	.9628	.9633	0	1	1	2	2	3	3	4	4
9.2	.9638	.9643	.9647	.9652	.9657	.9661	.9666	.9671	.9675	.9680	0	1	1	2	2	3	3	4	4
9.3	.9685	.9689	.9694	.9699	.9703	.9708	.9713	.9717	.9722	.9727	0	1	1	2	2	3	3	4	4
9.4	.9731	.9736	.9741	.9745	.9750	.9754	.9759	.9763	.9768	.9773	0	1	1	2	2	3	3	4	4
9.5	.9777	.9782	.9786	.9791	.9795	.9800	.9805	.9809	.9814	.9818	0	1	1	2	2	3	3	4	4
9.6	.9823	.9827	.9832	.9836	.9841	.9845	.9850	.9854	.9859	.9863	0	1	1	2	2	3	3	4	4
9.7	.9868	.9872	.9877	.9881	.9886	.9890	.9894	.9899	.9903	.9908	0	1	1	2	2	3	3	4	4
9.8	.9912	.9917	.9921	.9926	.9930	.9934	.9939	.9943	.9948	.9952	0	1	1	2	2	3	3	4	4
9.9	.9956	.9961	.9965	.9969	.9974	.9978	.9983	.9987	.9991	.9996	0	1	1	2	2	3	3	3	4

TABLE IX Values of the rank order correlation coefficient for various levels of significance

The probabilities given are for a one-sided test. For a two-sided test, the probabilities should be doubled.

n	r	P
4	1.000	.0417
5	1.000	.0083
5	.900	.0417
5	.800	.0667
5	.700	.1167
6	.943	.0083
6	.886	.0167
6	.829	.0292
6	.771	.0514
6	.657	.0875
7	.857	.0119
7	.786	.0240
7	.750	.0331
7	.714	.0440
7	.679	.0548
7	.643	.0694
7	.571	.1000
8	.810	.0108
8	.738	.0224
8	.690	.0331
8	.643	.0469
8	.619	.0550
8	.595	.0639
8	.524	.0956
9	.767	.0106
9	.700	.0210
9	.650	.0323
9	.617	.0417
9	.583	.0528
9	.550	.0656
9	.467	.1058
10	.733	.0100
10	.661	.0210
10	.612	.0324
10	.576	.0432
10	.552	.0515
10	.527	.0609
10	.442	.1021

SOURCE: The values of r were calculated from Table IV of E. G. Olds, Distributions of sums of squares of rank differences for small numbers of individuals. *Annals of Mathematical Statistics*, 1938, *9*, 133–148.

TABLE X Table of coefficients for orthogonal polynomials

k	Polynomial	Values of the Coefficients									
3	Linear	−1	0	1							
	Quadratic	1	−2	1							
4	Linear	−3	−1	1	3						
	Quadratic	1	−1	−1	1						
	Cubic	−1	3	−3	1						
5	Linear	−2	−1	0	1	2					
	Quadratic	2	−1	−2	−1	2					
	Cubic	−1	2	0	−2	1					
	Quartic	1	−4	6	−4	1					
6	Linear	−5	−3	−1	1	3	5				
	Quadratic	5	−1	−4	−4	−1	5				
	Cubic	−5	7	4	−4	−7	5				
	Quartic	1	−3	2	2	−3	1				
7	Linear	−3	−2	−1	0	1	2	3			
	Quadratic	5	0	−3	−4	−3	0	5			
	Cubic	−1	1	1	0	−1	−1	1			
	Quartic	3	−7	1	6	1	−7	3			
8	Linear	−7	−5	−3	−1	1	3	5	7		
	Quadratic	7	1	−3	−5	−5	−3	1	7		
	Cubic	−7	5	7	3	−3	−7	−5	7		
	Quartic	7	−13	−3	9	9	−3	−13	7		
9	Linear	−4	−3	−2	−1	0	1	2	3	4	
	Quadratic	28	7	−8	−17	−20	−17	−8	7	28	
	Cubic	−14	7	13	9	0	−9	−13	−7	14	
	Quartic	14	−21	−11	9	18	9	−11	−21	14	
10	Linear	−9	−7	−5	−3	−1	1	3	5	7	9
	Quadratic	6	2	−1	−3	−4	−4	−3	−1	2	6
	Cubic	−42	14	35	31	12	−12	−31	−35	−14	42
	Quartic	18	−22	−17	3	18	18	3	−17	−22	18

Index